SpringerBriefs in Geography

More information about this series at http://www.springer.com/series/10050

Wiesław Ziaja
Editor

Transformation of the Natural Environment in Western Sørkapp Land (Spitsbergen) since the 1980s

 Springer

Editor
Wiesław Ziaja
Institute of Geography and Spatial
   Management
Jagiellonian University
Cracow
Poland

First edition has been previously published by the Jagiellonian University Press in 2011
(ISBN: 978-83-233-3231-2): www.wuj.pl

ISSN 2211-4165   ISSN 2211-4173   (electronic)
SpringerBriefs in Geography
ISBN 978-3-319-26572-8   ISBN 978-3-319-26574-2   (eBook)
DOI 10.1007/978-3-319-26574-2

Library of Congress Control Number: 2015954621

Springer Cham Heidelberg New York Dordrecht London

Printed on acid-free paper

Springer International Publishing AG Switzerland is part of Springer Science+Business Media
(www.springer.com)

Acknowledgments

The book relies on the Research Project N N305 035634 *Changes in the western Sørkapp Land (Spitsbergen) natural environment due to global warming and human activity since 1982*, financed by the Polish Ministry of Science and Higher Education.

Contents

Book Review

This slim volume (92 pages) reports on approximately 25 years of landscape change in Sørkapp Land, the southern peninsula of Spitsbergen, the largest island of the Svalbard Archipelago.

Field research was conducted by a team of Polish scientists from Jagiellonian University, the legacy of an initial visit to the region by physical geographer Zdzislaw Czeppe during the International Geophysical Year 1957–1958. His interest in the research potential of the area was piqued, which led to a series of interdisciplinary summer expeditions beginning in 1980. The emphasis was on mapping abiotic and biotic features on a large scale (1:25,000–1:50,000). This resulted in a baseline of spatially detailed data that another team was able to repeat in an effort to detect change after another quarter century had passed.

Bruce Forbes, *Polar Record*, Vol. 51, Issue 3, 2015.

Editor and Contributors

About the Editor

Wiesław Ziaja is a physical geographer and landscape ecologist, a professor and head of the Department of Physical Geography, Institute of Geography and Spatial Management, Jagiellonian University in Cracow (Poland), a member of the International Arctic Science Committee Terrestrial Working Group and Committee on Polar Research, Polish Academy of Science. His research interests include landscape and natural environment structure and functioning, geographical aspects of nature and landscape protection, physical geography of the Arctic, North Europe, and the Carpathians. He is the author of papers for Polish and international journals, including *Arctic, Antarctic, and Alpine Research, Ambio, Annals of Glaciology, Polar Research,* and *Polish Polar Research.* He has broad experience of field expeditions to Spitsbergen.

Contributors

Justyna Dudek is a Ph.D. student at the Institute of Geography and Spatial Management, Jagiellonian University in Cracow, Poland. Her research interests concentrate on issues related to cold environments with special emphasis on the interaction between glaciers and landscapes, and current rapid environmental changes due to climate change in polar areas. She took part in the expeditions to Sørkapp Land in 2008 and 2010.

Maja Lisowska Ph.D. is a botanist with research interests in Arctic tundra vegetation changes, the role of cryptogams in primary succession, and European and Arctic lichens (biogeography, ecology, taxonomy, adaptations to extreme environmental conditions, air quality monitoring using lichens). Currently she has moved to science management, working as an Executive Director of Secretariat of the Polish Polar Consortium, and also at the Centre for Polar Studies, University of Silesia, Poland. In 2015, she joined the International Arctic Science Committee Secretariat as a coordinator of IASC Fellowships and early career support.

Maria Olech is a botanist, a professor of the Institute of Botany, Jagiellonian University in Cracow. Her main research activities concern taxonomy, ecology, and adaptations to extreme environmental conditions of high mountains and polar lichens, including the Arctic and Antarctic tundra vegetation dynamics with special regard to primary succession and biological colonization. She has broad experience of field expeditions to both the Arctic and Antarctic.

Krzysztof Ostafin Ph.D. is an associate professor in geographical information systems (GIS) at the Institute of Geography and Spatial Management, Jagiellonian University in Cracow. He specializes in landscape changes of mountainous areas, and took part in the expedition to Sørkapp Land in 2005. He is the coauthor of several papers in scientific journals (e.g., *Ambio, Land Use Policy, Global Environmental Change*) and books. He works with different sources of geographical data, for example, old maps and plans, repeat photographs, and satellite images. It is fascinating for him to merge this historical material with modern techniques such as GIS and environmental-landscape data that may be collected only in the field.

Piotr Osyczka is an associate professor, a lichenologist working at the Prof. Z. Czeppe Department of Polar Research and Documentation of the Institute of Botany, Jagiellonian University. His research interests are related to the various aspects of lichenology, concerning the lichens of polar regions and Poland. His scientific achievements are primarily associated with taxonomy, ecology, intraspecific variability, and environmental adaptations of lichens, their role in terrestrial ecosystems, and pioneer vegetation of cryptogams both in natural sites as well as anthropogenic and disturbed habitats. He is most involved in studies on the heavy-metal accumulation capacity of lichens and their role in spontaneous succession of strongly contaminated sites.

Michał Węgrzyn Ph.D. is a lichenologist and ecologist whose research interests are related to Arctic tundra vegetation. His main projects deal with phytosociology and vegetation mapping of the Arctic, as well as exploring phenomena that occur in the tundra under the influence of climate change. Currently he works at Prof. Z. Czeppe Department of Polar Research and Documentation of the Institute of Botany, Jagiellonian University in Cracow, Poland.

Chapter 1
Introduction: Study Area and Its Environmental Recognition

Wiesław Ziaja and Krzysztof Ostafin

Abstract Sørkapp Land, and especially its western part, has been chosen as a destination of Jagiellonian University scientific expeditions since 1980. Geographic locations and natural environmental features make this isolated and mountainous region unique in the European Arctic. This southern Spitsbergen peninsula constitutes a land wedge (which narrows to the south) between different seas. Its eastern coast is affected by the cold sea current, whereas the western one is under influence of warm Atlantic water. These factors help generate a great deal of variety in the peninsula's natural environment: from the western and southern lowlands overgrown by tundra with herds of reindeer to the glacial mountainous Arctic desert in the interior and east. Therefore, the peninsula is an excellent study area featuring all the relationships between the different components of the natural environment. They result in an unusually diverse, completely natural, and almost primeval landscape: glacial and periglacial, mountainous and low-lying, inland and coastal. These basic landscape types are internally differentiated: for example, fjord-type coastal landscape and open-ocean–type coastal landscape. Traces of the Pleistocene ice sheet may be discovered in the contemporary landscape. In addition, the reaction of this environment to climate warming can be readily noted due to relatively rapid climate fluctuations since the 1980s. New climate phenomena and associated trends can be easily observed in this area.

Keywords Sørkapp Land environmental differentiation · Jagiellonian University expeditions

W. Ziaja (✉) · K. Ostafin
Institute of Geography and Spatial Management, Jagiellonian University, Cracow, Poland
e-mail: wieslaw.ziaja@uj.edu.pl

© The Author(s) 2016
W. Ziaja (ed.), *Transformation of the Natural Environment in Western Sørkapp Land (Spitsbergen) since the 1980s*, SpringerBriefs in Geography, DOI 10.1007/978-3-319-26574-2_1

1.1 Course of the Landscape and Botanical Investigations in Western Sørkapp Land

Wiesław Ziaja

Zdzisław Czeppe, a physical geographer, viewed Sørkapp Land as a potential geographic area for scientific research. He was a participant in the Polish expedition during the Third International Geophysical Year, which wintered on the northern coast of Hornsund Fjord on Isbjørnhamna Bay in 1957–1958, and the summer expeditions of 1959 and 1960. Sørkapp Land, located south of the fjord, appeared to be an ideal place for research, which unfortunately could not be carried out at the time. However, he returned to western Sørkapp Land as a professor and the leader of Jagiellonian University expeditions in 1980 and 1981. He created a program of interdisciplinary research for this area, which was executed by the University's summer expeditions in the 1980s (the majority of them led by Adam Krawczyk). Landscape analysis played the most important role in the research of abiotic environmental features, whereas botanical analysis was crucial in the research of biotic features. The former was carried out in 1981–1984 and 1986, and the latter in 1982 and 1985. Six physical geographers (Z. Czeppe, P. Gębica, K. Kalicki, M. Kuczek, P. Libelt, and W. Ziaja) took part in field investigations of the landscape, and two botanists (E. Dubiel and M. Olech) took part in field investigations of vegetation. Their published results constitute the first relatively complete and reasonably detailed (maps at a scale of 1:25,000–1:50,000) characterization of the natural environment of the area.

The natural environment and landscape in Sørkapp Land have changed rapidly since the 1990s mainly due to climate warming as well as due to the progressive regeneration of fauna since the establishment of South Spitsbergen National Park in 1973. The effects of this transformation were observed during two short (two-week) Jagiellonian University summer scientific expeditions, which covered the northernmost part of western Sørkapp Land starting with the eastern coast of Gåshamna Bay: at the Konstantinovka hut in 2000 in the tent in 2005. An understanding of this transformation, in light of the accelerated evolution of the Arctic natural environment associated with global climate change, could help solve new and interesting research problems such as the regeneration of the reindeer population, which was virtually absent in the 1980s.

After a quarter of a century, two participants of the aforementioned expeditions in the 1980s, Maria Olech and Wiesław Ziaja, thought of repeating these investigations using similar methods but more advanced techniques (satellite remote sensing, GPS). Hence, together with younger colleagues, they proposed a research project titled "Changes in the Western Sørkapp Land Natural Environment due to Global Warming and Human Activity since 1982" and asked the Ministry of Science and Higher Education for financing. The project's objective was an analysis of changes in the structure and functioning of the western Sørkapp Land natural environment due to the aforementioned factors since the beginning of the 1980s. In 2008, the project was approved by the Ministry and a summer Jagiellonian University scientific expedition to western Sørkapp Land was organized.

The research team for this expedition was comprised of six people: Wiesław Ziaja (geographer, leader), for whom it was the ninth summer research season in Sørkapp Land and the twelfth in Spitsbergen, Piotr Osyczka (botanist), for whom it was the second summer season in Spitsbergen, Michał Węgrzyn (botanist), Justyna Dudek (geographer), Maja Lisowska (botanist), and Jan Niedźwiecki (geographer), who were spending their first summer in the Arctic. Five expedition members arrived in Longyearbyen by air on June 30th, 2008. Piotr Osyczka arrived via a research vessel operated by Gdynia Maritime University, *Horyzont II*. He arrived at the Polish Polar Station in Hornsund with the main expedition's baggage. After their arrival at Longyearbyen, the expedition members found shelter at Marek Zajączkowski's hospitable home. From Longyearbyen, a yacht called the *Eltanin*, under Jerzy Różański's (owner and shipmaster) command, transported them to the westernmost part of the Sørkapp Land coast. The first attempt (July 1–3) failed because of a dense icepack at the mouth of the Hornsund Fjord, and the yacht had to return to Longyearbyen. After a few days, three members of the expedition arrived from Longyearbyen to Hornsund via the *Eltanin* and the remaining members arrived via the *Horyzont II* on July 8th. The three expedition members who had arrived at the Polish Polar Station first, had boarded the yacht during its brief stop in Isbjørnhamna Bay, which is near the station. Immediately, the yacht sailed across the mouth of Hornsund Fjord to its southern shore in front of the trapper hut north of Palffyodden. Our colleagues from Hornsund Station, under Mateusz Moskalik's command, accompanied the yachts in motorboats loaded with the expedition's baggage (food, built materials, equipment). The expedition was conducted from the yacht to the Sørkapp Land coast with their help, landing on the coast on July 8th at 6:00 p.m. The staff of the Polish Polar Station in Hornsund, under the leadership of Marek Szymocha, helped our expedition in a number of ways, transporting our baggage and delivering necessary equipment several times. The geological team from the Polish Academy of Sciences in Warsaw, under the command of Krzysztof Krajewski, transported us via their dinghy to Breinesflya and Gåshamnøyra. Jerzy Czerny, our colleague from the AGH University of Sciences and Technology in Cracow, took one of the expedition members (who had to leave earlier from Spitsbergen) by boat from Sørkapp Land. On August 12th at 3:00 a.m., the expedition was transported by Krzysztof Krajewski's team from the coast near Palffyodden to the *Eltanin* yacht and departed from Sørkapp Land. We are very obliged to the aforementioned colleagues and their teams for their priceless help, which enabled us to realize all the elements of our plan. The expedition returned to Poland on August 15th.

1.2   Study Area

Wiesław Ziaja and Krzysztof Ostafin

Why is Sørkapp Land, and specifically western Sørkapp Land, such an attractive area for scientific research?

The answer is: Due to its geographic location and natural environmental features that make it unique in this part of the Arctic.

Sørkapp Land is the southern peninsula of Spitsbergen, the largest island of the Svalbard Archipelago. The island's east–west width is 150–200 km in its northern part. The island narrows to the south in the shape of a wedge between the Barents Sea in the east and the Greenland Sea to the west (Fig. 1.1). Their waters differ in temperature, which affects local climate conditions. The eastern coast is affected by the cold East Spitsbergen Current, which flows from the Arctic interior towards the south. This results in extensive glaciation and the lack of continuous vegetation. On the other hand, the western coast is warmed from the south by the West Spitsbergen Current, the last branch of the Gulf Stream. This produces relatively little glaciation and allows for a continuous tundra. However, southern winds cause a narrow belt of the cold current's waters to separate the warm current from the coast. Moreover, the two coasts differ in terms of geological structure and relief. In the east, the mountains are built of non-resistant Cretaceous and Tertiary rocks that fall down directly into the sea or onto narrow coastal lowlands (less than 1 km wide) at some locations. In the west, the mountains are separated from the sea by much wider (up to 4 km) coastal lowlands, with both cutting into hard Proterozoic and Paleozoic rocks. For this reason, the two Spitsbergen coasts are extremely different in terms of their natural environment and landscape (Hisdal 1985; Stange 2003). The farther to the south, the smaller the distance is between the two coasts.

This is true primarily of Sørkapp Land because the peninsula constitutes the end of the land wedge. Its maximum east–west width is 40 km in the north, and only 17 km in the south. Moreover, this wedge is "torn" in two places. In the north, two fjords, Hambergbukta from the east and Hornsund from the west, leave only a narrow isthmus between them. The width of the isthmus decreased to a mere 7 km in 2011 as a result of the present-day retreat of tidewater glacier fronts at the fjord heads due to the warming of the climate. The peninsula also gradually narrows in the south due to the retreat of the ice cliffs of the Vasilievbreen (east) and Olsokbreen (southwest) glaciers. The distance between the glaciers' cliffs has decreased to 12 km. The north–south length of the peninsula is 55 km (Fig. 1.2).

All of these factors help generate a great deal of variety in the peninsula's natural environment: from the western and southern lowlands overgrown by tundra with herds of reindeer to the glacial-mountainous Arctic desert in the interior and east.

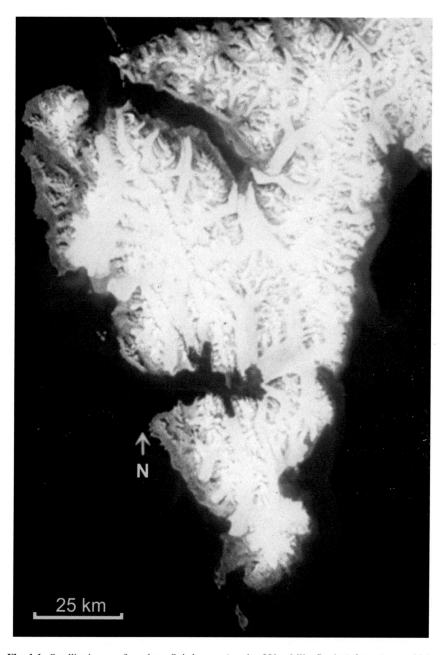

Fig. 1.1 Satellite image of southern Spitsbergen (south of Van Mijenfjorden) from August 20th, 1985: Landsat 5, MSS-421. Color scheme: *white*—glaciers; *red* or *reddish*—vegetation cover; *beige* and *blue-gray*—unglaciated areas without vegetation cover; *black*—seawater; *blue-and-black*—glacial water mixed with seawater

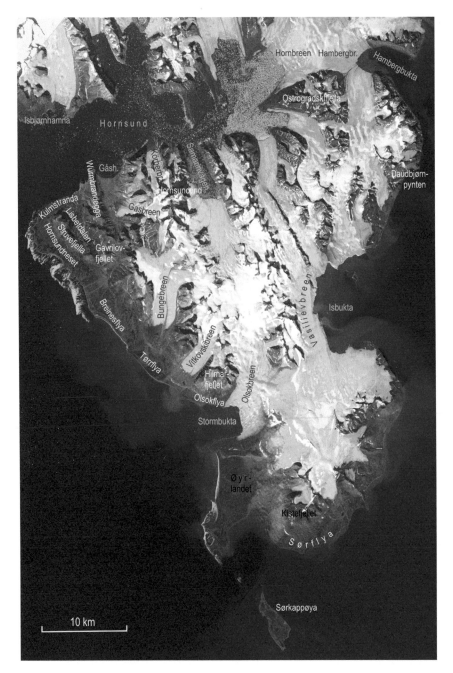

Fig. 1.2 Satellite image of Sørkapp Land consisting of four *TerraASTER* scenes from 2000 to 2004. Color scheme: *light-blue* with *white*—glaciers; *red* or *reddish*—vegetation cover; *brown*—land areas without vegetation cover; *dark-green* with lighter stripes—seawater with admixtures of glacial water. Most of the image is current for 2004

Therefore, the peninsula is an excellent study area featuring all the relationships between the different components of the natural environment. They result in an unusually diverse, completely natural, and almost primeval landscape: glacial and periglacial, mountainous and low-lying, inland and coastal. These basic landscape types are internally differentiated: for example, fjord-type coastal landscape and open-ocean–type coastal landscape. Traces of the Pleistocene ice sheet may be discovered in the contemporary landscape. In addition, the reaction of this environment to climate warming can be readily noted due to relatively rapid climate fluctuations since the 1980s. New climate phenomena and associated trends can be easily observed in this area.

Apart from that, Sørkapp Land belongs to the Arctic areas least accessible by sea, mainly because of wide shallows offshore, numerous submerged rocks, and relatively new bays and fjords at the tidewater glaciers' retreating fronts, which have not been surveyed yet.

Western Sørkapp Land (Fig. 1.3) is the most interesting part of the peninsula.

The area is accessible primarily by sea, as it protrudes between the open Greenland Sea and Hornsund Fjord, which results in the shortest season with sea-ice, that is, the longest period without a barrier to newcomers. Moreover, there is only one Sørkapp Land anchorage in Gåshamna Bay near the fjord's southern coast, close to its mouth.

Sørkapp Land was under the Barents Ice Sheet at least once during the Pleistocene, most likely during the Late Weichselian (Salvigsen and Elgersma 1993). The western part of the peninsula was completely covered by the ice sheet, which is evidenced by erratics on the Kovalevskifjellet peak at the elevation of 640 m (Ziaja 1989).

Most of western Sørkapp Land has been devoid of glaciers (Ziaja 1999) throughout the Holocene. The unglaciated area includes not only the coastal lowlands of Kulmstranda and Hornsundneset but also two mountain ranges: Struvefjella and a second unnamed range, which consists of Wurmbrandegga, Savitsjtoppen, Kovalevskifjellet, and Gavrilovfjellet. The unglaciated area also includes the Lisbetdalen valley between the two mountain ranges, which is an exception in Sørkapp Land and a rarity in Spitsbergen. Animals, plants, and soils are very common and fully visible in the area's landscape. This is not true in most parts of the peninsula. Little research had been conducted in the area until the 1980s. However, some hunting activity had been conducted there since the seventeenth century.

Animal life, plant life, and soils are also well developed on the low (elevation: below 50 m above sea level) coastal plains in the southern part of western Sørkapp Land, which is covered in part by tongues of glaciers extending from the peninsula's interior. Adjoining mountains are also partly glaciated on the land side. Two coastal plains, the wide Breinesflya plain at the foot of Wiederfjellet massif and the

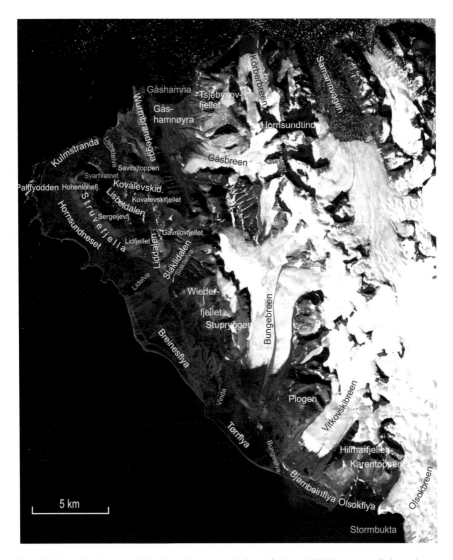

Fig. 1.3 Satellite image of Sørkapp Land consisting of *TerraASTER* scenes. Color scheme: *light-blue* with *white*—glaciers; *red* or *reddish*—vegetation cover; *brown*—land areas without vegetation cover; *dark-blue* with lighter stripes—seawater with admixtures of glacial water. The image is current for 2004

narrow Tørrflya plain at the forefield of the Bungebreen glacier, extend from the northwest to the southeast. Furthermore, narrow karst plains stretch at the base of the mountains from the Bungeelva proglacial river to the lateral marginal zone of the large Olsokbreen glacier.

The northernmost part of western Sørkapp Land consists of a low plain along Gåshamna Bay and the Tsjebysjovfjellet mountain massif, with a narrow lowland at its base on the fjord to the Körberbreen glacier in the east. This area, together with the Gåsbreen glacier, was investigated in detail in 1899 (De Geer 1923). Hence, the monitoring of changes in this glacier is the longest of all the more than 80 Sørkapp Land glaciers. The Gåsbreen glacier is a part of the boundary between the high-mountain–glacial interior and western part of the peninsula.

The lack of glaciers and a compact area of about 100 km^2 make this area relatively safe for exploration, which was particularly important for the early expeditions in the 1980s.

Animals in Sørkapp Land and the adjoining seas have been hunted since the seventeenth century. Historical remains including graves of Western European whalers and Pomor (North-Russian) hunters have been preserved in the western part of the peninsula.

The last Norwegian trapper station was operational until the establishment of South Spitsbergen National Park (south of Van Keulenfjord) in 1973 and has been preserved on the coast, 1 km north of the Palffyodden cape in an area most abundant in game. The station is a good base for summer scientific expeditions because its main hut is built of timber beams (two cosy rooms with stoves and a porch), making it resistant to polar bear attacks (Fig. 1.4: upper photo). A very small auxiliary hut (ca. 7 m^2, 1.5–1.7 m high), situated on the Breinesflya coast at a distance of about a half a day by foot from Palffyodden, also belongs to the former trapper station. The smaller hut, also built using timber beams, is a safe shelter for two individuals (Fig. 1.4: lower photo).

The three advantages of western Sørkapp Land—relatively easy access, lack of glaciers, and a former trapper station—are the more valuable due to the presence of many interesting environmental phenomena and historical sites.

In the 1980s, researchers went on foot from the huts to other parts of western Sørkapp Land. In most cases, they did not cross the Körberbreen glacier in the north and the Olsokbreen glacier in the south, as the glaciers flow towards the sea.

The glaciated peninsula's interior is extremely different from western Sørkapp Land because biotic features are rare and barely visible in the landscape of the peninsula interior. The boundary between the two regions is shifting slowly to the east due to the retreat of the glaciers.

Trapper hut near Palffyodden in 1986

Trapper shelter on Breinesflya in 2008

Fig. 1.4 The only standing huts in Western Sørkapp Land. *Photo* W. Ziaja

References

De Geer G (1923) Pl. F. Environs de la Station Russe D'Hivernage. 1:50,000 (map). In: Mesure D'un Arc de Méridien au Spitzberg, Entr. en 1899-1902, Description Topographique de la Région Explorée. Géologie. Aktiebolaget Centraltryckeriet, Stockholm

Hisdal V (1985) Geography of Svalbard. Polarhåndbook 2. Norsk Polarinstitutt, Oslo, p 81

Salvigsen O, Elgersma A (1993) Radiocarbon dating of deglaciation and raised beaches in north-western Sørkapp Land. Zeszyty Naukowe Uniwersytetu Jagiellońskiego, Prace Geograficzne 94:39–48

Stange R (2003) Rock and Ice. Landscape of the North. A geographical travelling accompaniment for Spitsbergen and East Greenland (68–74°N). Rolf Stange, 238 p

Ziaja W (1989) Rzeźba Doliny Lisbet i otaczających ją gór (Sörkappland, Spitsbergen). Zeszyty Naukowe Uniwersytetu Jagiellońskiego, Prace Geograficzne 73:85–97

Ziaja W (1999) Rozwój geosystemu Sørkapplandu, Svalbard. Wydawnictwo Uniwersytetu Jagiellońskiego, Kraków, 105 p

Chapter 2
Methods and Materials

Wiesław Ziaja, Justyna Dudek, Michał Węgrzyn, Maja Lisowska, Maria Olech and Piotr Osyczka

Abstract Landscape field mapping at a scale of 1:25,000 was the basic method of surveying the study area in the 1980s. Also the extent of the glaciers and their marginal zones as well as the sea coastline was mapped. The interpretation of aerial photographs, including the infrared ones, as well as the analysis of satellite data were combined with the results of field investigations. The following datasets were used in the manual delineation of the extent of each glacier in the chosen years: (1) three sheets of a topographic map 1:25,000, edited by the Polish Academy of Sciences in 1987, (2) a topographic map sheet at a scale of 1:100,000, edited by the Norwegian Polar Institute in 2007, (3) ASTER data with 15-m spatial resolution provided by NASA's Earth Observing System (EOS), and distributed by the Land Processes Distributed Active Archive Center (LPDAAC), and (4) SPOT 5 Orthophoto with 5-m spatial resolution, acquired on September 1st, 2008, provided by the IPY SPIRIT project. The maps were scanned and georeferenced to a common datum and projection. Changes in glacier surface elevation were estimated using digital elevation models. In the early 1980s, extensive vegetation research was carried out, using the Braun-Blanquet phytosociological method and a detailed vegetation map at a 1:25,000 scale was constructed. To evaluate the speed and direction of possible vegetation changes in the area, the aforementioned research was repeated in 2008. In both studies data were supported by photographs of the main vegetation types and their changes.

Keywords Landscape mapping · Phytosociological mapping · Remote sensing and GIS methods · Topographic maps · Aerial photographs · Satellite data

W. Ziaja (✉) · J. Dudek
Institute of Geography and Spatial Management, Jagiellonian University, Cracow, Poland
e-mail: wieslaw.ziaja@uj.edu.pl

M. Węgrzyn · M. Olech · P. Osyczka
Department of Polar Research and Documentation, Institute of Botany, Jagiellonian University, Cracow, Poland

M. Lisowska
Centre for Polar Studies, University of Silesia, Sosnowiec, Poland

© The Author(s) 2016
W. Ziaja (ed.), *Transformation of the Natural Environment in Western Sørkapp Land (Spitsbergen) since the 1980s*, SpringerBriefs in Geography,
DOI 10.1007/978-3-319-26574-2_2

2.1   Landscape Research

Wiesław Ziaja and Justyna Dudek

Complex landscape field mapping at a scale of 1:25,000 was the basic method of surveying the study area in the 1980s. A total of 1514 small basic landscape units (so-called geocomplexes of the *uroczysko* range) were mapped across an area of approximately 94 km^2. An old Norwegian topographic map at a scale of 1:100,000 (Sørkapp sheet), current for 1936 and enlarged fourfold, was used as a base map. Each individual landscape unit was drawn on the map, numbered, and described on a special separate form. The demarcation criteria consisted of spatial changes in environmental and landscape features. The following characteristics were analyzed: elevation, slope gradient and exposure, lithostratigraphy, tectonics (dip of the rock strata), Quaternary deposits (genesis, granulation, thickness), morphogenesis, microrelief, geomorphic processes, mesoclimate (based on exposure, shadowing, and persistence of snow patches), and bodies of water and vegetation (density and composition, division into vascular plants, mosses, and lichens). The mapping covered virtually all of the area that had not been covered by glaciers during the Holocene in western Sørkapp Land. The landscape units were grouped into several dozen types, which made it possible to create detailed maps of the landscape structure of the area (Czeppe and Ziaja 1985; Kuczek and Ziaja 1990; Ziaja 1991, 1992). This structure had not changed by 2008. However, certain changes within some of the landscape types were observed in 2008.

In addition, the extent of glaciers and their marginal zones as well as the sea coastline were mapped via a traditional method (using a Paulin altimeter) in the 1980s, and using more modern methods (GPS 12 Garmin) in 2008.

The interpretation of aerial photographs obtained from the Norsk Polarinstitutt, especially black-and-white photographs from 1961 and infrared photographs from 1990, as well as the analysis of satellite data were combined with the results of field investigations in a very effective way.

The analysis of glacier retreat since the 1980s was divided into two parts. The first part addressed changes in the glaciers' surface area in 1984, 1990, 2004, and 2007. In the second part, changes in the glaciers' surface elevation from 1990 to 2008 were inferred from two digital elevation models (see Maps in Chap. 4).

The following datasets were used in the manual delineation of the extent of each glacier:

1. Three sheets of a topographic map (Hornsund, Gåsbreen, and Bungebreen) at a scale of 1:25,000, edited by the Polish Academy of Sciences in 1987. The map was compiled photogrammetrically on the basis of vertical aerial photographs at a scale of 1:50,000 (taken in summer 1961) and geodetic measurements carried out during the Sixth Spitsbergen Expedition of the Academy during the 1983 and 1984 summer seasons.

2. A topographic map sheet (C13 Sørkapp) at a scale of 1:100,000, edited by the Norwegian Polar Institute in 2007. The map illustrates the extent of each glacier and provides contours at 50-m intervals. The part of the map with the study area was created based on vertical aerial photographs at a scale of 1:50,000 (taken in summer 1990).

3. ASTER data with 15-m spatial resolution provided by NASA's Earth Observing System (EOS), and distributed by the Land Processes Distributed Active Archive Center (LPDAAC): two consecutive scenes acquired on August 7th, 2004, and two consecutive scenes acquired on August 14th, 2007. All ASTER bands used in the study were orthorectified and projected to WGS84, UTM zone 33 north projection.

4. SPOT 5 Orthophoto with 5-m spatial resolution, acquired on September 1st, 2008. The data were provided by the IPY SPIRIT (International Polar Year SPOT5 stereoscopic survey of Polar Ice: Reference Images and Topographies) project (Korona et al. 2009). The mapping of glaciers was difficult in 2008 due to thick snow cover; therefore, only the frontal sections of tidewater glaciers and Bungebreen were updated using this dataset.

The maps were scanned and georeferenced to a common datum and projection (WGS84, UTM zone 33 north) using the four corners and 50 evenly distributed points for each map sheet edited by the Polish Academy of Sciences and about 100 points for the map edited by the Norwegian Polar Institute. The extent of each glacier based on Polish maps was validated and corrected using photographs taken in the field during Jagiellonian University expeditions in 1983 and 1984.

Changes in glacier surface elevation were estimated using two digital elevation models (DEMs). The first was a modern DEM with 20-m resolution compiled by the Norwegian Polar Institute from vertical aerial photographs at a scale of 1:50,000 taken in the summer 1990. In order to compare it to the second DEM, the dataset for 1990 was resampled to 40-m resolution. The first DEM was used as a reference dataset. The second DEM, with 40-m pixel size, was provided by the IPY SPIRIT project. It was generated from a cloud-free high-resolution SPOT 5 HRS image pair acquired on September 1st, 2008. The DEM extraction was performed by the French Institut Géographique National (IGN) following an automatic processing method, including manual intervention and interactive checks against ground-based measurements (Korona et al. 2009). The optical contrast of the images was mostly satisfactory, although in some parts, image matching had failed. Unmatched areas were masked and marked as "error in data". The SPIRIT DEM was co-registered using an analytical solution developed by Nuth and Kääb (2011), based on the cosinusoidal relationship between elevation differences and elevation derivatives of the slope and aspect. The method permitted minimizing a three-dimensional shift vector between two DEMs.

Changes in glacier elevation over the period 1990–2008 were estimated simply by subtracting the SPIRIT DEM from the reference DEM. The resulting values in the new raster indicated the difference in elevation at each pixel.

The work was performed in part at the Department of Geosciences, University of Oslo, with support granted by Iceland, Lichtenstein, and Norway by means of co-financing from the European Economic Area Financial Mechanism and the Norwegian Financial Mechanism as part of the Scholarship and Training Fund.

2.2   Botanical Research

Michał Węgrzyn, Maja Lisowska, Maria Olech and Piotr Osyczka

Western Sørkapp Land is diverse in terms of habitat conditions such as: topography, bedrock, hydrological and edaphic relationships, and microclimate. Lichens and bryophytes are predominant along with a few flowering plants, creating a complex mosaic of vegetation (Fig. 2.1). Therefore, the term plant community used herein applies both to vascular plants and cryptogamic organisms, that is, lichens and bryophytes.

During the 1982 and 1985 summer seasons, extensive vegetation research was carried out in northwest Sørkapp Land. The study area included marine terraces along the northern and western shores of Sørkapp Land, from the Lisbetelva river to the Vinda river, Hohenlohefjellet (614 m), as well as the western slopes of Sergeijevfjellet and Lidfjellet (Dubiel and Olech 1990).

Phytosociological research in the years 1982–1985 led to the identification of 28 vegetation units in the study area. The basis for the identification was 285 phytosociological relevés, performed according to the Braun-Blanquet (1964) method. Photographs were taken at different locations in order to obtain a full picture of the variety of vegetation. Complete phytosociological tables were created for 26 plant communities (Dubiel and Olech 1990), which were selected on the basis of their floristic characteristics. In some cases, the presence of dominant species strongly influencing the physiognomy of patches was considered more important. A short description of the habitat as well as comments on the distribution were added to the descriptions of vegetation. Names of differential species and their ecological scale were given for each community. Community names derive from the names of differential species, which were often dominant species. No new names of associations were created, as that would be premature, given the current state of knowledge of the diversity of the whole vegetation in Spitsbergen and the Arctic.

Next, field mapping was carried out. The arrangement of plant communities identified during phytosociological research was marked on a 1:25,000 map (Dubiel and Olech 1991). Vegetation units were marked on the map for four primary categories: (a) homogeneous plant communities occupying large areas; (b) examples of clear dominance of one community over others within a particular mosaic; (c) arrangement of two or three neighboring communities creating a mosaic; (d) large areas of transition between two or more communities.

In 2008 phytosociological research was repeated in the same plant communities, which had been identified in the 1980s (Dubiel and Olech 1990). The Braun-Blanquet

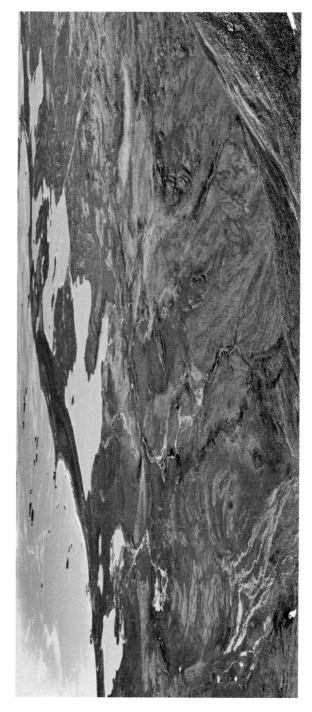

Fig. 2.1 Central part of the Hornsundneset terraced coastal plain built of Early Carboniferous sandstones, siltstones, and shales. A general view of the tundra vegetation and lakes (dammed by isostatically raised coastal ridges) from the slopes of Sergeijefjellet, facing WNW. *Photo* M. Węgrzyn, 2008

(1964) method was used once again. In phytosociological relevés, all groups of plants and fungi contributing to the composition of communities were included. The repetition of the phytosociological study in the same geographic area provided an accurate record of the state of the vegetation, and also allowed evaluating the speed and direction of changes. As was the case with earlier research (Dubiel and Olech 1990), the exact syntaxonomic position of certain plant communities was not precisely specified, inasmuch as it was not the aim of the research. However, the synonyms of the names of plants and fungi were updated.

In order to obtain data about changes in the size of plant communities, vegetation mapping was repeated on a 1:25,000 scale map. Field GPS devices containing a rectified version of the vegetation map from the 1980s, with areas of different vegetation units marked, were employed during the fieldwork. In many cases, the location of certain geomorphological elements such as streams, rivers, lakes, slopes, and shorelines, all of which create natural boundaries for plant communities, was adjusted as well.

Both vegetation maps were substantially improved. Relief was added to the maps, which created an opportunity for easier interpretation of the extent of each plant community. The final versions of both maps were prepared on the same scale of 1:50,000 in order to show vegetation changes more accurately (see Maps in Chap. 4).

Vegetation maps of the same geographic area created for different periods of time are the best way to show spatial changes in vegetation.

References

Braun-Blanquet J (1964) Pflanzensoziologie: Grundzüge der Vegetationskunde. 3. Auflage. Springer, New York, 865 p

Czeppe Z, Ziaja W (1985) Structure of the geographical environment of the north-western Sörkappland (Spitsbergen). Zeszyty Naukowe Uniwersytetu Jagiellońskiego, Prace Geograficzne 63:19–32

Dubiel E, Olech M (1990) Plant Communities of NW Sörkapp Land (Spitsbergen). Zeszyty Naukowe Uniwersytetu Jagiellońskiego, Prace Botaniczne 21:35–74

Dubiel E, Olech M (1991) Phytosociological map of NW Sörkapp Land (Spitsbergen). Zeszyty Naukowe Uniwersytetu Jagiellońskiego, Prace Botaniczne 22:47–54

Korona J, Berthier E, Bernard M, Remy F, Thouvenot E (2009) SPIRIT. SPOT 5 stereoscopic survey of Polar Ice: reference images and topographies during the fourth international polar year (2007–2009). ISPRS Journal of Photogrammetry and Remote Sensing 64:204–212

Kuczek M, Ziaja W (1990) The structure of the geographical environment of the area between two glaciers: Vitkovskibreen and Olsokbreen. Zeszyty Naukowe Uniwersytetu Jagiellońskiego, Prace Geograficzne 81:31–56

Nuth C, Kääb A (2011) Co-registration and bias corrections of satellite elevation data sets for quantifying glacier thickness change. Cryosphere 5:271–290

Ziaja W (1991) Fizycznogeograficzne zróżnicowanie górskiej części północno-zachodniego Sörkapplandu (Spitsbergen). Część I: Uroczyska. Zeszyty Naukowe Uniwersytetu Jagiellońskiego, Prace Geograficzne 83:31–54

Ziaja W (1992) Fizycznogeograficzne zróżnicowanie górskiej części północno-zachodniego Sörkapplandu (Spitsbergen). Część II: Tereny. Zeszyty Naukowe Uniwersytetu Jagiellońskiego, Prace Geograficzne 88:25–38

Chapter 3
Components of the Natural Environment

Wiesław Ziaja, Michał Węgrzyn, Maja Lisowska, Maria Olech and Piotr Osyczka

Abstract Western Sørkapp Land geological structure is very varied. Pre-Quaternary bedrock consists of three complexes: (1) the most extensive rocks from the Middle Proterozoic to the Silurian, folded in the Caledonian Orogeny: dolomites, phyllites, schists, quartzites, limestones, sandstones, breccias, and others; (2) Early Carboniferous clastic sediments: sandstones, including quartzitic sandstones, siltstones, and shales; and (3) Triassic sedimentary rocks: sandstones and conglomerates. None of these complexes lies horizontally. Loose Quaternary deposits (marine, glacial, fluvial, organic, frost-weathering, slope, including talus and solifluction) are not continuous. The climate is Arctic and marine-type. Average annual temperatures vary from −9 to −3 °C there. Average annual precipitation totals reach 300–500 mm. The mountains in northwest Sørkapp Land remain free of glaciers due to their exposure to relatively warm and dry eastern foehn winds. All the lowlands are unglaciated too. Sørkapp Land glaciation is clearly Arctic-type for two reasons: the common presence of permafrost, and the very weak influence of altitude on the distribution and extent of glaciers. The principal landform types in western Sørkapp Land are coastal lowlands, mountains, and mountain valleys. Nonglacial rivers and lakes (supplied directly by atmospheric precipitation and an active layer of permafrost) play an important part. There are also karst springs and glacial rivers and lakes in the northeast and southeast of the study area. Typically for the High Arctic, the flora of western Sørkapp Land is dominated by cryptogams, mainly lichens—about 170 species—whereas vascular plant flora includes 82 species. Different vegetation types often create complex mosaics, following diverse habitat conditions (bedrock, terrain relief, hydrology, etc.). In a few places the presence of seabird colonies has a local but strong impact on the vegetation.

W. Ziaja (✉)
Institute of Geography and Spatial Management, Jagiellonian University, Cracow, Poland
e-mail: wieslaw.ziaja@uj.edu.pl

M. Węgrzyn · M. Olech · P. Osyczka
Department of Polar Research and Documentation, Institute of Botany, Jagiellonian University, Cracow, Poland

M. Lisowska
Centre for Polar Studies, University of Silesia, Sosnowiec, Poland

© The Author(s) 2016
W. Ziaja (ed.), *Transformation of the Natural Environment in Western Sørkapp Land (Spitsbergen) since the 1980s*, SpringerBriefs in Geography, DOI 10.1007/978-3-319-26574-2_3

Keywords Bedrock · Climate · Glaciation · Landforms · Waters · Vegetation

3.1 Bedrock

Wiesław Ziaja

Solid pre-Quaternary bedrock in western Sørkapp Land can be grouped into three very different complexes:

1. The most extensive and oldest rocks, from the Middle Proterozoic to the Silurian, folded in the Caledonian Orogeny, formerly called Hecla Hoek, and Pre-Old Red basement today. This includes different types of dolomites, phyllites, schists, quartzites, limestones, sandstones, breccias, and others; between them, pre-Cambrian rocks are rather high-grade methamorphic rocks and Palaeozoic rocks are mainly low-grade methamorphic rocks;
2. Early Carboniferous clastic sediments in the northwest of the peninsula (protruding between the open Greenland Sea and Hornsund Fjord): sandstones, including quartzitic sandstones, siltstones, and shales (see Fig. 2.1);
3. Triassic sedimentary rocks that occur discordantly at some locations on the two aforementioned complexes: mainly sandstones and conglomerates.

None of these complexes lies horizontally. The rocks of the Pre-Old Red basement feature the largest dips: up to 90° and more (overturned folds). Dips of the next two complexes' rock strata tend to be much smaller: from 10 to 25°, sometimes more. All the complexes are cut by faults. The most important fault direction is NNW–SSE (Winsnes et al. 1992; Dallmann et al. 1993; Dallmann 1999).

Loose (unconsolidated) Quaternary deposits (marine, glacial, fluvial, organic, frost-weathering, slope, including talus and solifluction) are not continuous.

Coastal plains are partly covered by thin (up to 2–3 m), fragmented in many places, deposits of the Pleistocene and Holocene marine accumulation (rounded boulders and gravels, and sand). They are weathered to a substantial degree, apart from boulders and gravels of the Early Carboniferous quartzitic sandstone. Tørrflya, built of marine deposits up to approximately 10 m thick, is the exception.

Glacial deposits, both moraines and glacifluvial deposits, may be thicker in places (up to several meters), especially on coastal plains and mountain valley floors. Most are very young Holocene, from the Little Ice Age and the twentieth to twenty-first century (i.e., from the past several centuries), rarely from the end of the Pleistocene, such as numerous limestone (of the Slaklidalen formation) erratics.

Steep mountain slopes cut in solid rock are devoid of any superficial Quaternary deposits due to erosion-denudation processes which remove any regolith down. Talus and talus-torrent fans up to 10–20 m thick, in places connected into a one talus slope, lie at the bottoms of slopes. Less steep slopes (under 28–30°) can be overlain by weathering-solifluction deposits 1–2 m thick; the smaller the slope gradient, the more frequent and the thicker are the deposits.

Flat areas (0–3°) are mostly overlain by weathering (on top plateaus) or accumulation types of cover, which can be as much as 2 m thick. Flat areas undergo further weathering and frost segregation. A thin (up to 30 cm) soil–vegetation cover is deposited across flat areas.

A unique superficial microrelief is a feature of each type of Quaternary deposit.

3.2 Climate

Wiesław Ziaja

The climate in Sørkapp Land and throughout Spitsbergen is Arctic and marine-type. Average annual temperatures vary from −9 to −3 °C in the west of Sørkapp Land; that is, the temperatures are quite high in relation to the latitude of almost 77° N, due to the warm West Spitsbergen Current. Average summer (July and August) temperatures are usually in the range of 3–5 °C, thus they are relatively low due to the ocean's influence. Average annual precipitation totals are usually in the range of 300–500 mm. Average summer (July and August) precipitation totals are much more variable and range from 30 to 200 mm in different years. Since the beginning of the twentieth century, no yearly meteorological observations have been carried out in western Sørkapp Land. Hence, the aforementioned climate characteristics were based on meteorological observations carried out at the station on Isbjørnhamna Bay, which belongs to the Institute of Geophysics of the Polish Academy of Sciences (Marsz and Styszyńska 2013). The station is located north of Hornsund Fjord at a 10-km distance from the northern coast of western Sørkapp Land. The data from this station are compatible with the results of meteorological observations carried out at different locations in western Sørkapp Land during selected summer seasons in the twentieth century (Ziaja 1999).

Climate fluctuations during the twentieth century are very important in the context of this book's subject matter, especially the last fluctuation, which has occurred since the 1980s and is described below in Chap. 4 on environmental changes.

The climate in western Sørkapp Land is unique in terms of the entire peninsula and southern Spitsbergen because its features indicate a lack of glaciers in the local mountains throughout the Holocene. These mountains are not lower than mountains in the rest of the peninsula, which are more or less covered by glaciers. The mountains in northwest Sørkapp Land remain free of glaciers due to their exposure to relatively warm and dry eastern foehn winds, which melt the winter snow cover. This occurs not only today but has been occurring since the beginning of the Holocene, even during periods of climate cooling. The lack of glaciers in the Holocene has been shown by geological (Salvigsen and Elgersma 1993) and geomorphological (Ziaja 1989) investigations. The foehn effect was detected on the basis of meteorological observations (Kalicki 1985; Ziaja 1985).

3.3   Glaciation

Wiesław Ziaja

Sørkapp Land glaciation is clearly Artic-type for the following reasons:

1. Common presence of permafrost, which is predominantly frozen solid rock and only in the Quaternary deposits consists of ice mixed with rock material of variable granulation;
2. Very weak influence of altitude (above sea level) on the distribution and extent of glaciers.

The northern part of western Sørkapp Land (Fig. 1.3) includes about 50 km^2 of mountains (seven massifs arranged into two ridges with the Lisbetdalen valley between them) without glaciers during all of the Holocene, in spite of the fact that their quite extensive and flattened peaks reach an elevation of 640 m (Ziaja 1992, 1999). This is due to the exceptional local climate described earlier (equally high or lower areas situated a few kilometers farther to the east and southeast have been superficially glaciated to a substantial degree).

These areas with expansive glaciers flowing from the glaciated peninsula's interior to the west and south, and with small glaciers on coastal mountain massifs, are not part of western Sørkapp Land. The extent of each glacier marks the natural boundary of the western region. A retreat of this extent is simply a shift of this boundary to the east.

The coastal lowlands—which stretch to the west of the Körberbreen glacier on Hornsund Fjord and farther to the southeast to the Olsokbreen glacier on the open Greenland Sea—were also unglaciated throughout the Holocene. The unglaciated slopes of mountains falling to the lowlands of the peninsula's middle west (Breinesflya and others to the southeast) may also be thought of as western Sørkapp Land. However, these mountains are more or less glaciated from the land (interior) side.

Thermo-mineral springs indicate the existence of a talik, which is not covered by a glacier or lake in the southernmost corner of western Sørkapp Land (in Bjørnbeinflya and Olsokflya), analogically to what has been described by Salvigsen and Elgersma (1985).

3.4   Terrain Relief

Wiesław Ziaja

The principal landform types in western Sørkapp Land are coastal lowlands, mountains, and mountain valleys (Fig. 1.3).

Coastal lowlands constitute more than half of the study area and were formed as a shallow seabed offshore during the Pleistocene and the Holocene, and uplifted due to isostatic movements afterwards.

There are two coastal lowlands along Hornsund Fjord, which are isolated from other lowlands by the slopes of the Tsjebysjovfjellet and Wurmbrandegga massifs running down to the sea in the northeastern part of the study area. The narrow (up to 250 m wide) and short (ca. 1.5 km long) lowland with the Stonehengesteinane group of rocks lies at the foot of the northern wall of Tsjebysjovfjellet called Rasstupet. The second lowland, Gåshamnøyra, consists of the virtually flat outlet of a wide valley covered by extensive (2 km × 2 km) extramarginal sandur formed by proglacial waters of Gåsbreen Glacier.

The Kulmstranda lowland (1.5–2.0 km wide), built of resistant Early Carboniferous sandstones, adjoins from the south of the outlet of Hornsund Fjord. The lowland is formed of extensive rock terraces, which are covered by a thin layer of quartzitic marine pebbles or devoid of them in many places. The lowland's higher part forms a ridge reaching 123 m. The Lisbetdalen valley is located behind this ridge. Kulstranda is incised by the gorge of the Lisbetelva river (ca. 10 m deep). The second, more shallow and already inactive, incision in the lowland is situated 300 m farther to the east. This lowland experiences extremely strong and frequent winds that blow away fine-grained rock material and limit plant life.

The Hornsundneset lowland, 1.5–3.0 km wide, with the hut near Palffyodden, stretches from the fjord's outlet towards the south and southeast. Successive (ever older) marine terraces are formed in tiers from the coastline up to the foot of the Struvefjella mountain range, at an altitude of 5–70 m (Fig. 2.1). As was the case in Kulmstranda, the terraces also cut into resistant Early Carboniferous quartzitic sandstones. The sandstones' small dip into the land side brings out the terraces' edges, and their rock surfaces are mostly covered by old coastal ridges built of marine pebbles. The lowland is fertilized by small auks, *Alle alle*, that nest in the mountain slopes above it, which helps promote vegetation and soil formation.

Breinesflya, the largest of the Sørkapp Land lowlands, is completely different from the aforementioned lowlands: 3–4 km wide, boggy, gradually rising from the beach to the foot of Wiederfjellet massif at an elevation of approximately 50 m, built of less resistant Triassic sedimentary rocks. The lowland is almost completely covered by a thick (most often 1–2 m) layer of fine-grained marine sediments along with an admixture of material washed away from the slopes above them as well as fluvial material ranging from clay to debris. This type of cover favors vegetation and soil formation (Fig. 3.1).

Three narrow (up to 1 km wide) lowlands lie farther to the southeast. Tørrflya is the accumulation marine terrace up to 10 m high and thick, with the cliff coast on Triassic bedrock, between the formerly major Vinda river and the large Bungeelva glacial river. The extra-marginal sandurs of Bungebreen Glacier adjoin Tørrflya. The next lowland, Bjørnbeinflya, stretches to the southeast of the Bungeelva river. This lowland is at first narrower between the Vitkovskibreen glacier's marginal zone and the sea, and then wider at the foot of Hilmarfjellet, featuring erosion karst forms such as holes and underground channels. The next lowland, Olsokflya, is located along the Stormbukta bay. Thin Triassic layers overlie the limestone Pre-Old Red basement only along the coastal belt up to 300 m wide.

Fig. 3.1 Northwestern Sørkapp Land seen from the Breinesflya low and wet coastal plain to the northwest. The following mountains are visible in the background (from *left* to *right*): Hohenlohefjellet and Sergeijevfjellet (joined together), Lidfjellet, Gavrilovfjellet, and the northwestern fragment of Wiederfjellet. *Photo* J. Niedźwiecki, 2008

The large (5 km × 4 km) Stormbukta Bay did not exist at the end of the Little Ice Age in 1900 (Wassiliew 1925). The only tidewater glacier found along the western Sørkapp Land coast, Olsokbreen, covered the bay's present-day area. At the time, the glacier was much thicker and larger than it is today. Its northern (right) lateral moraine has changed into a wide marginal zone due to the glacier's retreat. A new coastline has also appeared in the area.

All the coastal lowlands are shaped by weathering and frost segregation, which can be observed in the form of characteristic microrelief.

Mountains dominate the landscape despite the fact that they occupy less than half of western Sørkapp Land. The largest quantities of water and deposits flow to the lowlands and the sea from the mountains due to higher precipitation and intensive erosion-denudation processes taking place therein.

Struvefjella, the westernmost mountain range, consists of three different massifs (Fig. 3.1). Hohenlohefjellet (Fig. 3.2), the northernmost and the highest (>600 m) of the massifs, cuts into resistant Lower Carboniferous sandstones. It is primarily covered by quartzitic debris, which creeps gravitationally (especially on the eastern slopes) or reaches its angle of repose (especially on other slopes). High nivation moraines, which often overlie old glacial moraines, and talus and talus-torrent fans are situated at the foot of the slope and in slope gullies. High nivation moraines are built of coarse sandstone blocks. Sergeijevfjellet is a ridge 3 km long and built of less-resistant Lower Carboniferous and Triassic sedimentary rocks, thus lower (ca.

400 m) and with more gentle slopes (up to 30°) under a weathering–solifluction cover. The high (>500 m) Lidfjellet massif is shaped like a truncated pyramid built of Mesozoic sedimentary rocks. Its steep slopes primarily undergo weathering and gravitational creeping. Its top is shaped like a triangle with a regular inclination gradient of about 10°.

The Struvefjella range is dissected by two narrow, short (1.0–1.5 km), and deep (altitude: below 145 m) tectonic valleys: Hohenloheskardet (Fig. 3.2) and Sergeijevskardet, shaped by the Barents Ice Sheet and the sea during the Pleistocene as well as slope processes during the Holocene. There is a Young Pleistocene glacial moraine ridge at the mouth of the Sergeijevskardet. There are also numerous limestone (Slaklidalen formation) erratics in the valleys and on the slopes of Sergeijevfjellet and Lidfjellet.

The Lisbetdalen mountain valley (Fig. 3.2), located east of Struvefjella, is expansive and cuts deep (altitude: below 180 m) into resistant Lower Carboniferous sedimentary rock. Traces of the presence of the Pleistocene ice sheet (deep basin of Svartvatnet Lake, limestone erratics) and the sea (marine gravels) can be observed in this area. The valley was shaped during the Holocene by weathering, solifluction, sheet wash, nivation, and fluvial processes. The lateral Kovalevskidalen falls into the Lisbetdalen from the east.

The second unglaciated (and unnamed) mountain range in western Sørkapp Land towers over the Lisbetdalen from the east. The range consists of the following four mountain massifs: the narrow and steep Wurmbrandegga ridge built of Proterozoic dolomites (ca. 400 m), and farther to the south, the Savitsjtoppen (almost 500 m), the Kovalevskifjellet (640 m), and the Gavrilovfjellet (almost 600 m) massifs. Gentler western slopes and top plateaus of the last three massifs are primarily built of Lower Carboniferous and Triassic sedimentary rocks and have been shaped by geomorphic processes similar to those in Lisbetdalen (apart from fluvial processes). The steep eastern slopes of these three massifs, built of non-resistant Proterozoic phyllites, undergo intensive weathering and creeping. These slopes fall down to two valleys, which form the boundary between western and central Sørkapp Land: the northern unnamed valley with the Gåsbreen glacier's marginal zone and the Slaklidalen valley, which runs to the south.

Limestone erratics can be found on the top plateau of Kovalevskifjellet. Hence, this mountain range must have been covered by the ice sheet at the end of the Pleistocene (Ziaja 1989; Salvigsen and Elgersma 1993).

The deep and narrow V-shaped Liddalen Valley can be found at the junction of the two aforementioned ranges (Sruvefjella and the unnamed range), south of Lisbetdalen, between Gavrilovfjellet and Lidfjellet.

The long (6 km) and high (>700 m) Wiederfjellet-Stupryggen mountain range stretches to the southeast from Slaklidalen to the Bungebreen glacier. This range, wide in the northwest and narrow in the southeast, is built of different types of rocks of the Pre-Old Red Basement. The slopes of the range's northwestern part are cut in the aforementioned phyllites and undergo weathering, sheet wash, and solifluction. The rest of the range, built of dolomitic sandstones and metamorphic limestones, is extremely asymmetric. Its gentle eastern slopes are covered partly by the

Fig. 3.2 Lisbetdalen valley with Svartvatnet lake seen from the east. In the background: the eastern slopes of Hohenlohefjellet, which fall down to the lake and the northeastern slopes of Sergeijevfjellet. Large talus-torrent fans and nival moraines are visible at the base of Hohenlohefjellet on the lake. *Photo* W. Ziaja, 2008

Wiederfjellet glacier. Its steep western slopes (weathering and creeping) fall down to the bend (on the fault) between the range and the Breinesflya coastal plain. A narrow belt of gentle solifluction slopes has formed on this bend.

The Plogen massif, almost 700 m high and with steep slopes, is located between the valleys of the Bungebreen and Vitkovskibreen glaciers. The Hilmarfjellet-Karentoppen massif (>800 m) is located in the southeastern corner of the study area between the Vitkovskibreen and Olsokbreen glaciers. Both massifs are built of metamorphic limestones and sandstones of the Pre-Old Red Basement, and their peaks are built of Triassic sedimentary rocks. The highest marine raised terraces in Spitsbergen (up to 338 m) are preserved on the slopes of Hilmarfjellet.

The last three mountain massifs are at least partly glaciated from the land side. However, their slopes, which fall down to the coastal lowlands, are free of glaciers.

3.5 Waters

Wiesław Ziaja

Karst springs can be found at the foot of Tsjebysjovfjellet on Hornsund Fjord (Pulina 1977; Leszkiewicz 1982) in the northeastern corner of the study area. Gåshamnøyra plain, a large extramarginal sandur located west of the massif, was formed by streams from both Gåsbreen Glacier and the valley south of the glacier's front edge. These streams flow across the glacier's marginal zone built mainly of dead ice undergoing rapid melting.

Non-glacial rivers and lakes play an important part in the western Sørkapp Land water system. This is not true elsewhere in Spitsbergen. The rivers and lakes in western Sørkapp Land are supplied directly by atmospheric precipitation—including snow patches, which thaw before the following winter—and an active layer of permafrost. This is especially true of the northern part of the study area due to a complete lack of glaciers there. The largest body of fresh and clean water is the Svartvatnet lake, which is found in a deep glacial erosion trough. The lake's water table is found at an altitude of 72 m above sea level. The lake contains a population of *Salvelinus alpinus* L. salmonoid fish (Gullestad and Klemsten 1997). Two streams flow into the lake: a main stream, which flows from the snow patch located in the pass (at almost 200 m) between the Lidfjellet and Kovalevskifjellet massifs, and a lateral stream flowing from the Kovalevskidalen valley. The entire Lisbetdalen valley is drained by the Listbetelva, which is the only permanent river in the northern part of the study area, flowing across the Kulmstranda plain. The next river, the Lidelva, flows from the Liddalen valley and along the southern edge of Hornsundneset plain to the sea. The two rivers are up to 0.5 m deep. Several small streams located between the two rivers flow from the slopes of the Struvefjella across the coastal plains to the sea. Their gradient is relatively large (ca. 100 m per 1.5–

3.0 km on the plains). Both rivers are mostly dry, as they consist of rocky terraces with thin and fragmented marine gravel cover. Numerous shallow lakes (up to 1 m deep), permanent near the coast and seasonal closer to the mountains, have been created due to the damming of water by new and old raised coastal ridges, often found on the sandstone layers that dip towards the interior of the land (Fig. 2.1). The exceptions are two small lakes in glacial troughs: Savitsjvatnet Lake on the highest part of Kulmstranda plain and an unnamed lake with fish in the pass between the Sergeijevfjellet and Lidfjellet massifs.

The Slaklielva river flows through the Slaklidalen valley located along the eastern boundary of the study area. The river splits up into two streams at the valley outlet and flows to the sea along the northern edge of Breinesflya plain, approximately 1 km south of Lidelva. Wiederfjellet mountain slopes are drained by other streams, which flow across the plain. Only the Slaklielva stream and a stream flowing from Wiederdalen are supplied partly by small glaciers. All the streams are supplied by the active layer of permafrost because the plain is covered by thick fine-grained deposits. Numerous swamps and the slow outflow of water are caused by the small gradient of the plain (50 m per 3–4 km) and large areas of moss tundra (Fig. 3.1). The rivers are up to 1 m deep. There are several small lakes located more than 1.5 km from the sea in the southern part of the plain.

The wide braided channel of the Vinda glacial river cuts off the Breinesflya plain from the southeast. The Vinda river flows out of the Bungebreen glacier's marginal zone. The largest river in western Sørkapp Land, the Bungeelva, flows out of the Bungebreen glacier to the south and across its marginal zone and drains the Tørrflya coastal plain from the east.

There are several surprisingly small superficial streams, first of all proglacial streams, across the plain's narrow neck between the Vitkovskibreen glacier's front moraine and the sea (0.5–1.0 km) behind the Bungeelva river. There are virtually no streams on the karst lowlands of Bjørnbeiflya and Olsokflya.

Water from the Vitkovskibreen glacier and the Hilmarfjellet massif flow via underground karst channels across the plain to the sea (Pulina 1977). There is a very large karst spring (several m^3/s), Trollosen, in the outlet of the biggest channel on the sea-coast (Fig. 3.3).

There are two thermo-mineral springs, up to 16.5 °C, with H_2S and CO_2 (Werenskiöld 1920; Liestøl 1976; Pulina 1977; Krawczyk and Pulina 1991): in Bjørnbeinflya, where they supply a few small and very shallow lakes, and in Olsokflya. The small Luktvatnet lake (which means smelly), near the Olsokbreen's lateral moraine, may be supplied in a similar manner.

There are few melt-water lakes and, closer to the glacier, proglacial streams in the Olsokbreen glacier's extensive marginal zone, in the southeastern corner of the study area.

Fig. 3.3 Outlet of a large underground karst channel on the Olsokflya plain coast with the Trollosen karst spring and one of the karst holes (1 m in diameter, enlarged in the upper photo and marked with a red arrow in the lower photo) that developed above the channel after 1986 as a result of intensifying karst processes. *Photo* J. Dudek, 2008

3.6 Vegetation

Michał Węgrzyn, Maja Lisowska, Maria Olech and Piotr Osyczka

The vegetation within Svalbard is dominated by cryptograms, mainly lichenised fungi (lichens), circa 600 species, and bryophytes, circa 370 species (Elvebakk and Hertel 1996), and procariotic and eucariotic algae (Matuła et al. 2007). Vascular plants are in the minority with about 170 species (Jónsdóttir 2005). Similar proportions can be observed in western Sørkapp Land where lichen biota consists of 170 species (Olech 1990), and vascular plant flora includes 82 species (Dubiel 1990). This is typical for the High Arctic. Western Sørkapp Land is diverse in terms of habitat conditions such as shape, geological subsoil, hydrological conditions, edaphic relationships, microclimate, and more. Vegetation in the region is dominated by lichens and bryophytes, and some flowering plants. Different vegetation types often create a complex mosaic.

The key factors determining the location of different types of vegetation in the Arctic areas are terrain relief and hydrological conditions (Dubiel and Olech 1990; Elvebakk 1997).

Bedrock has an obvious influence on vegetation. The presence of carboniferous sandstone correlates with the location of epilithic lichen communities and moss communities with *Racomitrium lanuginosum*, and lichen tundra with *Flavocetraria nivalis* and *Cladonia rangiferina*. Triassic formations, in turn, overlap with vegetation dominated by vascular plants, for example, *Saxifraga nivalis*, *Bistorta vivipara*, and *Papaver dahlianum*. Other communities such as the *Gymnomitrion coralloides* community, the *Cetrariella delisei* community, and the *Bistorta vivipara* community, are found on coastal ridges and small hills.

One of the most important environmental factors determining the structure and location of tundra vegetation is the availability of water in the substratum (Dubiel and Olech 1990; Elvebakk 1997). In extremely dry places, where melting of snow is the principal source of water, and water is quickly removed, the following communities of epilithic lichens can be found: the *Orphniospora moriopsis* community, the *Umbilicaria arctica* community, and the *Umbilicaria cylindrica* community. In addition, the lichen-dominated *Flavocetraria nivalis–Cladonia rangiferina* community and the *Cetrariella delisei* community are also found.

In depressions, often supplied with water by seasonal and summer-active streams, mire and wetland vegetation can be found. This includes grass and bryophyte-dominated communities such as the *Calliergon stramineum* community, the *Juncus biglumis* community, the *Dupontia pelligera–Arctophila fulva* community, the *Carex subspathacea* community, and the *Saxifraga hyperborea–Ranunculus spitsbergensis* community.

Communities of ornithocoprophilous vegetation develop on mountain slopes or close to mountain slopes as well as in the vicinity of large seabird colonies and in places visited by birds. These include the communities with *Phippsia algida*, the bryophyte-dominated *Tetraplodon mnioides* community as well as the community of crustose lichens with *Candelariella arctica*.

There has been a tendency over the last few years to create more general classifications of plant communities and to present them on general maps. Two general vegetation maps containing information about the study area are available: the Circumpolar Arctic Vegetation Map (CAVM Team 2003), and the Vegetation Map of Svalbard (Elvebakk 2005). According to the Circumpolar Arctic Vegetation Map, northwest Sørkapp Land is located in bioclimatic Subzone B, the second-coldest of the five subzones in the Arctic (CAVM Team 2003). Vegetation in Subzone B is generally open or patchy (20–80 % cover), with plants 5–10 cm tall, and a relatively high (30–60 %) proportion of lichens and mosses, and this characteristic matches the vegetation in the study area quite well. Dry sites in northwest Sørkapp Land are dominated by the zonal vegetation type of Subzone B, "prostrate dwarf-shrub, herb tundra" (P1 type). Another tundra-type characteristic for Subzone B, "sedge/grass and moss wetlands" (W1 type), is dominant at wet sites. A more detailed vegetation map of Svalbard by Elvebakk (2005) shows three major vegetation units in northwest Sørkapp Land: mesic, circumneutral tundra characterized by *Luzula nivalis*, mesic, acidic tundra characterized by *Luzula confusa* and *Deschampsia alpina* mires.

In polar terrestrial ecosystems, animals play a major role in the circulation of matter and in the flow of energy. Seabirds nesting in coastal areas are a particularly

Fig. 3.4 A herd of reindeer grazing on the slopes of Sergeijevfjellet. Damaged vegetation and a major change in slope microrelief (appearance of so-called cattle terraces due to intensive trampling) are visible. *Photo* M. Węgrzyn, 2008

important part of Arctic ecosystems, as their presence determines the functioning of terrestrial biocenoses. Seabirds link terrestrial and marine ecosystems, as they feed in the sea and breed on land, and thus bring a considerable amount of organic matter from sea to land, mainly in the form of guano. It is a very important source of nutrients for tundra vegetation, as an allochthonous source of biogenic substances, especially nitrogen and phosphorus, which are in short supply. Nutrient enrichment of areas adjacent to seabird colonies, which causes stimulation of primary production and development of luxuriant vegetation, is called the ornithocoprophilic effect (Eurola and Hakala 1977). Organic compounds found in guano are easily absorbed by plants.

Over the last few years, vegetation in the tundra of Sørkapp Land has been greatly influenced by herds of reindeer, which appeared in the 1990s. Reindeer forage on vascular plants, mainly graminoids and fruticose lichens, is causing significant changes in vegetation (Fig. 3.4).

References

CAVM Team (2003) Circumpolar Arctic Vegetation Map. Scale 1:7,500,000. Conservation of Arctic Flora and Fauna (CAFF) Map No. 1. U.S. Fish and Wildlife Service, Anchorage, Alaska

Dallmann WK (ed) (1999) Lithostratigraphic Lexicon of Svalbard. Upper Palaeozoic to Quaternary bedrock. Review and recommendations for nomenclature use. Norsk Polarinstitutt, Tromsø, p 318

Dallmann WK, Birkenmajer K, Hjelle A, Mørk A, Ohta Y, Salvigsen O, Winsnes TS (1993) Geological Map, Svalbard, 1:100,000, C13G, Sørkapp (text and map). Norsk Polarinstitutt Temakart 17:73

Dubiel E (1990) Vascular plants of the NW Sörkapp Land (Spitsbergen). Distribution and habitats. Zeszyty Naukowe Uniwersytetu Jagiellońskiego, Prace Botaniczne 21:7–33

Dubiel E, Olech M (1990) Plant Communities of NW Sörkapp Land (Spitsbergen). Zeszyty Naukowe Uniwersytetu Jagiellońskiego, Prace Botaniczne 21:35–74

Elvebakk A (1997) Tundra diversity and ecological characteristics of Svalbard. In: Wielgolaski FE (ed) Polar and alpine tundra. Ecosystems of the World, 3. Elsevier, Amsterdam, pp 347–359

Elvebakk A (2005) A vegetation map of Svalbard on the scale 1:3,5 mill. Phytocoenologia 35 (4):951–967

Elvebakk A, Hertel H (1996): Part 6 Lichens. In: Elvebakk A, Prestrud P (eds) A catalogue of Svalbard plants, fungi, algae and cyanobacteria. Norsk Polarinstutt Skrifter 198:271–359

Eurola S, Hakala AVK (1977) The bird cliff vegetation of Svalbard. Aquilo, Serie Botanica 15:1–18

Gullestad N, Klemsten A (1997) Size, age and spawning frequency of landlocked arctic charr Salvelinus alpinus (L.) in Svartvatnet, Svalbard. Polar Res 16(2):85–92

Jónsdóttir IS (2005) Terrestrial Ecosystems on Svalbard: heterogeneity, complexity and fragility from an Arctic Island perspective. Biol Environ Proc R Ir Acad 105(3):155–165

Kalicki T (1985) The foehnic effects of the NE winds in Palffyodden region (Sörkappland). Zeszyty Naukowe Uniwersytetu Jagiellońskiego, Prace Geograficzne 63:99–106

Krawczyk WE, Pulina M (1991) Thermokarstic and glaciokarstic hydrology in Spitsbergen glaciers. In: Proceedings, 1st international symposium of Glacier Caves and Karst in Polar Regions, Madrid, p 185–198

Leszkiewicz J (1982) Źródła krasowe pod Rasstupet na południowym Spitsbergenie w świetle badań z 1978 roku. Wyprawy Polarne Uniwersytetu Śląskiego 1977–1980, tom I: 77–86

Liestøl O (1976) Pingos, springs, and permafrost in Spitsbergen. Norsk Polarinstitutt årbok 1975:7–29

Marsz AA, Styszyńska A (eds) (2013) Climate and climate change at Hornsund, Svalbard. Gdynia Maritime Uniwersity, Gdynia, p 402

Matuła J, Pietryka M, Richter D, Wojtuń B (2007) Cyanoprokaryota and algae of Arctic terrestrial ecosytems in the Hornsund area. Spitsbergen. Pol Polar Res 28(4):277–282

Olech M (1990) Lichens of the NW Sörkapp Land (Spitsbergen). Zeszyty Naukowe Uniwersytetu Jagiellońskiego, Prace Botaniczne 21:197–210

Pulina M (1977) Uwagi o zjawiskach krasowych w południowej części Spitsbergenu. Kras i speleologia 1(X): 104–129

Salvigsen O, Elgersma A (1985) Large-scale karst features and open taliks at Vardeborgsletta, outer Isfjorden, Svalbard. Polar Res 3:145–153

Salvigsen O, Elgersma A (1993) Radiocarbon dating of deglaciation and raised beaches in north-western Sørkapp Land. Zeszyty Naukowe Uniwersytetu Jagiellońskiego, Prace Geograficzne 94:39–48

Wassiliew (Vasiliev) AS (1925) Spitsberg III. Feuille du Sud. (map). In: Missions scientifiques pour la mesure d'un arc de méridien au Spitsberg. Miss. Russe, Publication de l'Académie des Scientes de Russie

Werenskiöld W (1920) Spitsbergens fysiske geografi. Naturen (Bergen) 44:209–242

Winsnes TS, Birkenmajer K, Dallmann WK, Hjelle A, Salvigsen O (1992) Geological Map, Svalbard, 1:100000, Sørkapp. Norsk Polarinstitutt Temakart, 17

Ziaja W (1985) The influence of winds on the temperature and humidity at north-western Sörkappland (Spitsbergen) in summer 1982. Zeszyty Naukowe Uniwersytetu Jagiellońskiego, Prace Geograficzne 63:107–113

Ziaja W (1989) Rzeźba Doliny Lisbet i otaczających ją gór (Sörkappland, Spitsbergen). Zeszyty Naukowe Uniwersytetu Jagiellońskiego, Prace Geograficzne 88:25–38

Ziaja W (1992) Fizycznogeograficzne zróżnicowanie górskiej części północno-zachodniego Sørkapplandu (Spitsbergen). Część II: Tereny. Zeszyty Naukowe Uniwersytetu Jagiellońskiego, Prace Geograficzne 73:85–97

Ziaja W (1999) Rozwój geosystemu Sørkapplandu, Svalbard. Wydawnictwo Uniwersytetu Jagiellońskiego, Kraków, p 105

Chapter 4
Environmental and Landscape Changes

Wiesław Ziaja, Justyna Dudek, Krzysztof Ostafin, Michał Węgrzyn, Maja Lisowska, Maria Olech and Piotr Osyczka

Abstract Climate changes in western Sørkapp Land mirror global fluctuations. The Little Ice Age ended with a cold period in the 1890s. A warm contemporary period began in the early twentieth century. Afterwards, secondary cold and warm climate fluctuations occurred. The most recent fluctuation, since the 1980s, shows a significant warming trend. The mean annual temperature increased by almost 2 °C and the mean annual total precipitation increased by about 60 mm since the 1980s (according to data of the station located 10 km from the study area). Almost all the snow patches melt during the warmest and sunniest summer seasons. The so-called active layer of permafrost had doubled at sites below 100 m of altitude from the 1980s to 2008. Almost all Sørkapp Land glaciers have undergone a continuous recession since the beginning of the twentieth century. Two processes are important for glacier recession: decrease in snow accumulation in firn fields due to the summer thawing of a larger snow mass, and summer thawing of ice on the surface of glacier tongues, which results in a decrease in ice thickness. Thus, the equilibrium-line altitude of a glacier shifts upward, reducing the accumulation zone. Hence, the entire surface of the glaciers undergoes lowering each year, which results in a decrease in their volume and their overall retreat. Since the 1980s, an acceleration of the glaciers' recession has occurred, causing great changes in landforms and Quaternary. New accumulation landforms appeared in the front of glaciers and around glacier tongues in their marginal zones, that is, on lowlands and valley floors abandoned by glaciers and in their forefields situated below the marginal zones (i.e., beyond the former extent of the glaciers). New erosion landforms, apart from proglacial river incisions, prevail on the steep slopes of valleys and mountain massifs. The cliffs of tidewater glaciers undergo the quickest retreat. Karst processes have intensified due

W. Ziaja (✉) · J. Dudek · K. Ostafin
Institute of Geography and Spatial Management, Jagiellonian University, Cracow, Poland
e-mail: wieslaw.ziaja@uj.edu.pl

M. Węgrzyn · M. Olech · P. Osyczka
Department of Polar Research and Documentation, Institute of Botany, Jagiellonian University, Cracow, Poland

M. Lisowska
Centre for Polar Studies, University of Silesia, Sosnowiec, Poland

W. Ziaja (ed.), *Transformation of the Natural Environment in Western Sørkapp Land (Spitsbergen) since the 1980s*, SpringerBriefs in Geography, DOI 10.1007/978-3-319-26574-2_4

to higher air temperatures and larger quantities of flowing water. Surface and underground streams carry much more water today than in the 1980s. However, the soil is generally drier on the lowlands between the streams today due to the deepening of the active layer above the permafrost. The river and lake network changed the most due to glacier recession. Ice-dammed lakes disappeared due to the recession of glaciers. On the basis of repeated vegetation mapping, significant changes in composition and extent of several plant communities were documented. Decrease in species diversity, leading to a more uniform vegetation, has been observed mainly in dry habitats. In some cases boundaries between plant communities that were clear in the 1980s have now vanished. Fruticose epigeic lichens, such as *Flavocetraria nivalis*, *Cladonia rangiferina*, and other species of *Cladonia* have disappeared from the most part of the study area and their extent is now limited to steep rocky slopes. In some communities increase in abundance of *Salix polaris* was recorded. The main cause of vegetation changes in Sørkapp Land is the rapidly growing reindeer population in the area.

Keywords Climate warming · Glacial recession · Transformations of landforms · Changes in water bodies · Changes in vegetation

4.1 Climate Warming

Wiesław Ziaja

Changes in the climate in western Sørkapp Land mirror global climate fluctuations. This is especially true of the northern hemisphere and Spitsbergen. There is a lack of meteorological data to show continuous climate fluctuations for the entire Sørkapp Land peninsula. Indeed, basic meteorological observations (mainly temperature) were carried out in different places throughout the peninsula but for very short periods of time, mostly in the summer (Ziaja 1999). Continuous meteorological observations have been carried out at the Polish Polar Station, which belongs to the Institute of Geophysics of the Polish Academy of Sciences, on the Isbjørnhamna bay (Marsz and Styszyńska 2013). The bay is located along the northern coast of Hornsund Fjord, only 10 km from the study area. Hence, the station's data may be considered representative of western Sørkapp Land.

The Little Ice Age ended with a very cold period in the 1890s in Spitsbergen, including Sørkapp Land. A relatively warm contemporary period "after the Little Ice Age" began in the early twentieth century. Secondary cold and warm climate fluctuations occurred during the twentieth and early twenty-first century. The most recent fluctuations, starting in the 1980s, appear to point to a significant warming trend (Brázdil 1988; Ziaja 2004). This is certainly true of the western island's coast, both north and south of Hornsund Fjord. The mean annual temperature increased by almost 2 °C and the mean summer (July and August) temperature increased by 0.6 ° C since the 1980s when we began our research (Table 4.1).

Table 4.1 Mean annual temperature changes at the Polish Polar Station at Horsund, according to data from the Institute of Geophysics, Polish Academy of Sciences, published in part by Styszyńska (2013)

	1980–1989	2000–2009
Mean annual temperature (°C)	–5.2	–3.15
Mean summer (i.e., July and August) temperature (°C)	+3.9	+4.5

Table 4.2 Mean annual changes in precipitation totals at the Polish Polar Station at Horsund, according to data from the Institute of Geophysics, Polish Academy of Sciences, published in part by Styszyńska (2013)

	1980–1989	2000–2009
Mean annual sum of precipitation (mm)	390.5	451.3
Mean summer (i.e., July and August) sum of precipitation (mm)	78.1	94.4

A systematic temperature increase occurred during each month. The greatest warming occurred during a winter season from September to February. This statistically significant positive trend in the change in air temperature, also in June and August (Marsz 2013), accelerated the thawing of snow and ice. This contributed to the lengthening of the vegetation season, the thawing of the permafrost active layer, and the ablation of glaciers.

An increase in precipitation, both annual totals (statistically significant trend) and summer (July and August) totals has been detected since the 1980s. A clear increase in rain precipitation totals, and the intensity and frequency of rainfall have been noted during this time period. This is also true of the summer season (Łupikasza 2013; Table 4.2). This increased rainfall intensified a number of geomorphic processes as well as snow and ice ablation. This, in turn, impacted the water cycle and vegetation.

Marsz and Styszyńska (2013) argue that the climate in southwestern Spitsbergen has changed rapidly and considerably during the past 30 years. The following are symptoms of this change: warming, an increase in oceanicity in terms of air temperature, an increase in precipitation totals, increasing share of rain in precipitation totals, a shortening of the snow season, and an increase in cloudiness. This climate change prompted other changes as well.

4.2 Glacial Recession

Wiesław Ziaja and Justyna Dudek

Snow patches, permafrost, and glaciers are the environmental components most sensitive to climate warming.

All the snow patches melt during the warmest and sunniest summer seasons across western Sørkapp Land lowlands and mountains, except for the patch where the Lisbetelva river originates. In spite of climate warming, there are still summer seasons with numerous snow patches that last until the next winter season.

Permafrost is more isolated from the summer heat (positive temperatures) by a layer of ground, which thaws down to a certain depth, that is, the so-called active layer. In spite of the lack of a detailed survey, an increase in the active layer's thickness has been observed. Its maximum thickness in Quaternary deposits (apart from blocks and boulders) varied between 40 and 1.5 m, according to their granulation and vegetation, at sites below 100 m of altitude in 1986 (Ziaja 1988), and had doubled by 2008. Rapid melting of thin (up to 2 m) ice lenticels occurs in rare pingo landforms in the lowest marine terrace (Fig. 4.1).

Glaciers react to warming with a few years (or longer) delay. Almost all Sørkapp Land glaciers have undergone a continuous recession since the beginning of the twentieth century. Only three small glaciers in the highest part of the Hornsundtind mountain group increased in thickness (and volume) from the 1930s to the 1980s (Ziaja 1999), but then became smaller as well. Two small glaciers, on the eastern slopes of Kovalevskifjellet and western slopes of Wiedefjellet, had already declined before the 1980s (Ziaja 1999). A few surges (i.e., rapid ice flows) resulted in temporary (up to several years) advances of a glacier terminus or in the thickening of a glacier tongue, and could have occurred in western Sørkapp Land during the past century. The Bungebreen glacier's surge after 1990 (Fig. 4.2) was described by Sund et al. (2009). Surges during a period of climate warming lead to the glaciers' recession because a downward shift of an ice mass results in the thinning of glaciers' firn fields, and thickened glacier tongues melt quickly, reducing the glaciers' volume and surface area.

Fig. 4.1 Destruction of a small pingo (several m long, up to 1 m thick) on the accumulation coast south of Palffyodden. *Photo* J. Dudek, 2008

Fig. 4.2 Fragment of the Bungebreen glacier front seen (cross-section) from the east, with characteristic ice stratification following the glacier's surge. *Photo* J. Dudek, 2010

The following processes are important for glacier recession in the study area: (1) a decrease in snow accumulation in firn fields due to the thawing of a larger snow mass in the summer, and (2) summer thawing of ice on the surface of glacier tongues, which results in a decrease in ice thickness in the tongues. Thus, the equilibrium-line altitude between the ablation zone and the accumulation zone of a glacier shifts upward due to warming, which reduces the accumulation zone. Hence, the entire surface of the glaciers undergoes lowering each year, which results in a decrease in their volume and their overall retreat. The retreat occurs everywhere and not just at the front of the glacier. The thinning of glaciers around nunataks and in valleys with steep slopes causes the uncovering of rocks or rock-and-weathering slopes from underneath the ice. Large areas of slopes are visually reduced using a normal projection on a map. Frontal glacier retreat, which leads to the appearance of marginal zones, is most noticeable, as glaciers are least inclined in the front. However, freeing marginal zones of glaciers need not mean freeing them of ice. Marginal zones are predominantly formed out of dead (motionless) glacial ice covered with a thin (up to 2 m) and continuous layer of moraine or glacifluvial material. This dead ice successively thaws during the summer season. However, the thawing process is so slow that the volume of ice is still (after more than 100 years of glacial recession) larger than the volume of new moraine or glacifluvial deposits. Of course, landform and Quaternary deposits develop over time, as described next.

During the period of our study, since the 1980s, an acceleration of the glaciers' recession had occurred (Maps 1a–1b and 2a–2b, Fig. 4.3). Why? First, because the glaciers in question are too large in relation to contemporary climate conditions and

(a)

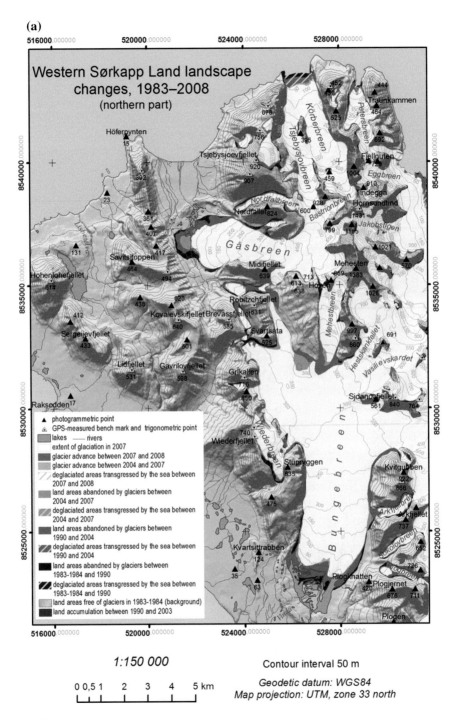

Western Sørkapp Land landscape
changes, 1983–2008
(northern part)

▲ photogrammetric point

△ GPS-measured bench mark and trigonometric point

lakes —— rivers

extent of glaciation in 2007

glacier advance between 2007 and 2008

glacier advance between 2004 and 2007

deglaciated areas transgressed by the sea between 2007 and 2008

land areas abandoned by glaciers between 2004 and 2007

deglaciated areas transgressed by the sea between 2004 and 2007

land areas abandoned by glaciers between 1990 and 2004

deglaciated areas transgressed by the sea between 1990 and 2004

land areas abandned by glaciers between 1983-1984 and 1990

deglaciated areas transgressed by the sea between 1983-1984 and 1990

land areas free of glaciers in 1983-1984 (background)

land accumulation between 1990 and 2003

1:150 000

Contour interval 50 m

0 0,5 1   2   3   4   5 km

Geodetic datum: WGS84
Map projection: UTM, zone 33 north

Map 1a

(b)

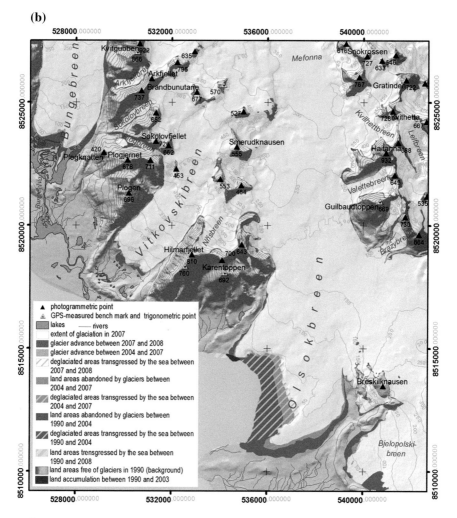

Both maps (northern and southern part):
Contour lines on areas free of glaciers derive from topographic map (C13 Sørkapp) at a scale of 1:100 000 (made on
the basis of vertical aerial photographs taken in 1990), edited by the Norwegian Polar Institute in 2007.
Outlines of glaciers derive from the following sources:
- 3 map sheets (Hornsund, Gåsbreen, Bungebreen) at a scale of 1:25 000, edited by the Polish Academy of Science
 in 1987 and field observations carried out by the Jagiellonian University expeditions in 1982-1984 (only northern part);
- topographic map (C13 Sørkapp) at a scale of 1:100 000, edited by the Norwegian Polar Institute in 2007;
- ASTER data provided by NASA's Earth Observing System (EOS), available from the Land Processes Distributed
Active Archive Center: two consecutive scenes acquired on August 7th, 2004 and two consecutive scenes acquired on
August 14th, 2007;
- front extents of tidewater glaciers and Bungebreen in 2008 as well as contour lines on glacieted areas were updated
from SPIRIT Program material © CNES 2009 and Spot Image 2008, all rights reserved.

This work was in parts conducted at the Department of Geosciences, University of Oslo, with the support granted by
Iceland, Liechtenstein and Norway by means of co-financing from the European Economic Area Financial Mechanism
and the Norwegian Financial Mechanism as part of the Scholarship and Training Fund.

Maps include results of field observations carried out by the Jagiellonian University expedition in summer 2008,
financed by the Polish Ministry of Sciences and Higher Education (project N N305 035634).

Map 1b

(a)

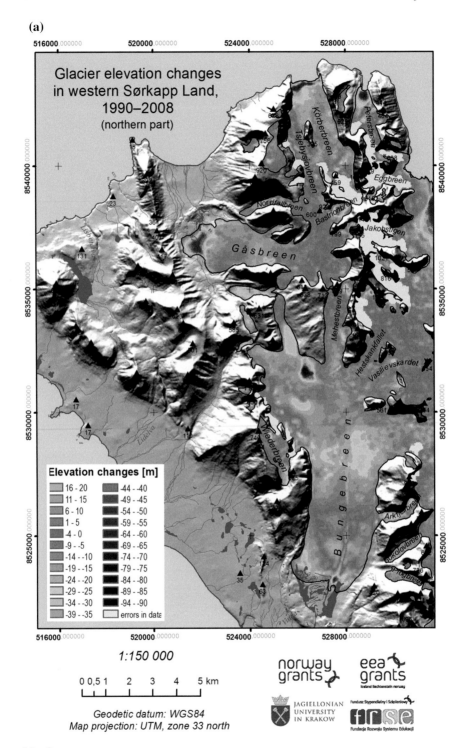

Map 2a

(b)

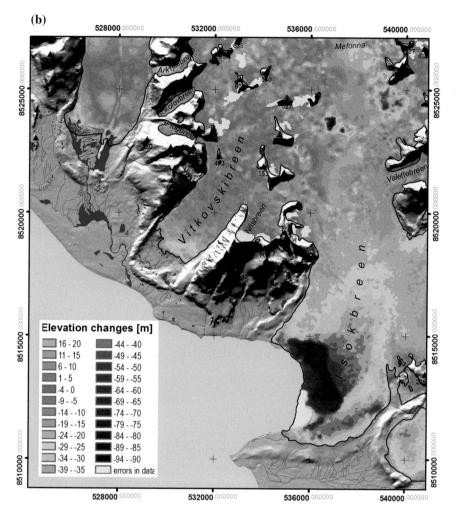

Both maps (northern and southern part):
Glaciers elevation changes derive from the following data sets:
1) a digital elevation model (DEM) with pixel spacing 40 m generated by the Norwegian Polar Institute from vertical aerial photographs taken in 1990.
2) a digital elevation model (DEM) with pixel spacing 40 m provided by SPIRIT Program © CNES 2009 and Spot Image 2008, all rights reserved.

This work was in parts conducted at the Department of Geosciences, University of Oslo, with the support granted by Iceland, Liechtenstein and Norway by means of co-financing from the European Economic Area Financial Mechanism and the Norwegian Financial Mechanism as part of the Scholarship and Training Fund.

Map 2b

Fig. 4.3 Landscape changes in the Gåshamna area: the Goësvatnet glacial-dammed lake (which still existed in 2000) disappeared before 2005 because of the glacier's recession. The Gåsbreen glacier is much shorter now and does not dam up the valley. View from the south. *Upper photo* W. Ziaja. *Lower photo* J. Niedźwiecki

especially in relation to air temperature. Thus, the glaciers would shrink even after an end to climate warming. Second, since the 1980s, significant warming has additionally stimulated the set of processes described above and positive feedback as well. New factors stimulating glacier recession have appeared, for example, an increase in the share of rain in annual precipitation totals.

The glaciers' thickness generally decreased from 1990 to 2008, usually by 10–40 m, and in places, especially in the lower parts and edges of the glaciers, as much as 40–94 m. Olsokbreen, the largest glacier in terms of area and volume, has experienced the largest decrease in the thickness of ice (Fig. 4.4). A large decrease in the thickness of the upper parts of the glaciers is almost certain because the summer season of 2008 was exceptionally cloudy and not windy, and the winter snow mass did not melt as usual. This can be easily shown by comparing *TerraASTER* satellite images from 2007 and from 2008. Only in the lower part of the Bungebreen glacier did its thickness increase by up to 20 m due to a surge (Maps 2a–2b).

If one defines western Sørkapp Land as the region west of the glacial interior, the recession of the glaciers leads to the expansion of western Sørkapp Land. The marginal zones of large glaciers—Gåsbreen (Figs. 4.3 and 4.5), Bungebreen (Fig. 4.2), and Vitkovskibreen (Fig. 4.6)—have become considerably wider since the 1980s. The retreat or decline of the smaller glaciers between them has occurred as well. The recession of the glaciers changed how the environment and the landscape type of the two valleys along the eastern boundary of the study area

Fig. 4.4 Inner Stormbukta bay with the new shoreline formed since the 1980s, seen from the slopes of Hilmarfjellet (altitude: 335 m). *Photo* J. Dudek, 2008

Fig. 4.5 Landscape changes along a boundary between western Sørkapp Land and the peninsula's interior (with the Hornsundtind mountain group, 1431 m, in the background): the Gåsbreen glacier's recession resulted in (1) great changes in terrain relief in valleys, (2) disappearance of the Goësvatnet glacial-dammed lake, and (3) formation of a new terraced river valley. View from the Savitsjtoppen mountain massif to the east. *Upper photo* Pillewizer (1939). *Lower photo* J. Dudek

function. The unnamed valley in the north, which falls towards Gåshamna Bay, is at this point not dammed by the Gåsbreen glacier (and was still dammed in 2000) and does not feature an ice-dammed lake (which existed in 2000). This glacier has become thinner and its front has retreated 750 m since the 1980s (Fig. 4.3). Hence, the valley has become easily accessible: it is possible to follow its floor from the coast of Gåshamna to the Slaklidalen valley and farther to the south of the study area, because even the smaller glaciers in the area had disappeared or became smaller along with numerous snow patches. The front of the Wiederbreen glacier (on the gentle eastern slopes of Widerfjellet) has retreated far enough to become invisible from the Breinesflya lowland. A significant retreat of Bungebreen Glacier (almost 1 km) has occurred since the 1980s (Map 1a–1b, Fig. 4.2), in spite of the surge after 1990 (Sund et al. 2009). The low-situated

Fig. 4.6 Vitkovskibreen glacier's front and marginal zone seen from the slopes of Hilmarfjellet (altitude: 330 m). View from the southeast. *Photo* J. Dudek, 2008

tongue of this glacier became thicker temporarily due to the surge. The frontal retreat of Vitkovskibreen glacier (Fig. 4.6) was much slower: 300–400 m since 1990 (Map 1b). The glaciers were concave and ended close (300–400 m) to the sea at the beginning of the twentieth century, and then underwent a successive but moderate decrease in thickness.

The tidewater Olsokbreen glacier, which marks the end of the study area from the southeast, has undergone the largest (in the study area) recession after the Little Ice Age. A very large decrease in its thickness has resulted primarily in its frontal retreat, which caused Stormbukta (the largest bay along the study area coast) to appear. The Olsokbreen front has retreated 1–2 km since 1990 (Map 1b, Fig. 4.4). The recession of the tidewater Körberbreen glacier, which marks the end of the study area from the northeast, is also noticeable but the degree of recession is much smaller (Map 1a) due to its higher elevation and northern exposure.

Map 1a covers western Sørkapp Land without the vicinity of the Vitkovskibreen and Olsokbreen glaciers but with the western part of the peninsula's interior. In the area found on Map 1a, the rate of the surface recession of glaciers was very similar in the 1980s and 1990s: 0.66–0.71 km^2, that is, 0.6–0.7 % of glaciated area per year. This rate has decreased somewhat since 2004. The share of the glaciated area covered by Map 1a decreased from almost 34 % (106 km^2) in 1982–1984 to almost

Table 4.3 Changes in glaciation in the northwestern part of Sørkapp Land, covered approximately by Map 1a

Year	Land area (km²)	Glaciated area (km² %)	Unglaciated area (km² %)	Land area decrease (km² %)	Rate of glacial recession (km² % of the glaciated area per year)	Areas abandoned by glaciers above sea level (km² %)	Deglaciated areas transgressed by the sea (km² %)
1983	313.85	106.30 **33.87**	207.55 **66.12**	0.27 **0.09**	0.66 **0.60**	4.37 **94.18**	0.27 **5.82**
1990	313.59	101.66 **32.41**	211.92 **67.59**				
1990	313.58	101.66 **32.41**	211.37 **67.59**	0.55 **0.18**	0.72 **0.71**	8.76 **93.99**	0.56 **6.01**
2004	313.04	92.34 **29.50**	220.71 **70.50**				
2004	313.04	92.34 **29.50**	220.71 **70.50**	0.09 **0.03**	0.46 **0.50**	1.69 **92.86**	0.13 **7.14**
2007	312.95	90.52 **28.93**	222.42 **71.07**				
1983	313.85	106.30 **33.87**	207.55 **66.12**	0.9 **0.30**	0.66 **0.60**	14.82 **93.82**	0.96 **6.08**
2007	312.95	90.52 **28.93**	222.42 **71.07**				

29 % (90.5 km²) in 2007. A very small part (only 6 %) of the area abandoned by glaciers was inundated by the waters of Hornsund Fjord: almost 1 km², that is, 0.3 % of the area covered by Map 1a (Table 4.3).

During the period 2004–2007 and the period 2007–2008, the very small areas abandoned by glaciers (0.219 km² in sum) became occupied by the Körberbreen and Bungebreen glaciers, which generally retreat rather quickly (Map 1a). This episode did not affect the overall retreat of glaciers in western Sørkapp Land.

Map 1b covers the southeastern part of western Sørkapp Land in the vicinity of the Vitkovskibreen and Olsokbreen glaciers. In the area found on Map 1b, the rate of the surface recession of glaciers between 1990 and 2007 was slightly slower than that in the northern part of the study area, that is, 0.5 % of the glaciated area per year. However, the surface area in this case is actually much larger. This rate decreased to some extent after 2004, impacted by the same short-term climate fluctuations affecting the entire study area. The share of the glaciated area covered by Map 1b decreased from 61 % (169 km²) in 1990 to almost 57 % (154.3 km²) in 2007. A substantial part (almost 30 %) of the ice-free areas was transgressed by the waters of the Stormbukta bay: 4.3 km², that is, almost 2 % of the area covered by Map 1b (Table 4.4).

Table **4.4** Changes in glaciation in the southwestern part of Sørkapp Land, covered approximately by Map 1b

Year	Land area (km^2)	Glaciated area (km^2 %)	Unglaciated area (km^2 %)	Land area decrease (km^2 %)	Rate of glacial recession (km^2 % of glaciated area per year)	Areas abandoned by glaciers above sea level (km^2 %)	Deglaciated areas transgressed by the sea (km^2 %)
1990	276.76	168.94 **61.04**	107.82 **38.96**	3.32 **1.20**	0.89 **0.52**	9.38 **74.03**	3.14 **25.07**
2004	273.43	156.42 **57.20**	117.00 **42.79**				
2004	273.43	156.42 **57.20**	117.00 **42.79**	1.30 **0.48**	0.70 **0.45**	0.94 **44.76**	1.16 **55.24**
2007	272.13	154.32 **56.71**	117.81 **43.29**				
1990	276.76	168.94 **61.04**	107.82 **38.96**	4.63 **1.68**	0.86 **0.51**	10.32 **70.59**	4.30 **29.41**
2007	272.13	154.32 **56.71**	117.81 **43.29**				

4.3   Transformation of Landforms and Quaternary Deposits

Wiesław Ziaja, Krzysztof Ostafin and Justyna Dudek

The greatest changes in landforms and Quaternary deposits have occurred due to the glaciers' recession, as described above. New landforms have appeared. Accumulation landforms appeared in the front of the glaciers and around glacier tongues in their marginal zones, that is, on lowlands and valley floors abandoned by glaciers and in their forefields situated below the marginal zones (i.e., beyond the former reach of the glaciers, Figs. 4.2, 4.3, 4.4, 4.5, 4.6). Erosion landforms, apart from proglacial river incisions (Fig. 4.7), prevail on the steep slopes of valleys and mountain massifs. Hence, the recession of the upper parts of glaciers results almost exclusively in erosion landforms, and the recession of the lower parts of glaciers leaves both erosion and accumulation landforms behind (with predominance of the latter)

The following new ice-free landforms have been observed: rocky and rock-and-weathering slopes, valley incisions (cut by glacial waters) of different types (gullies, gorges), and roche moutonnées. The decrease in glacier thickness uncovers nunatak slopes from top to bottom. The continuation of this process leads to the joining of nunataks into new mountain ridges or ranges. Once the ice is gone, these erosion landforms undergo mostly nonglacial denudation-erosion changes,

Fig. 4.7 Wide bed of the Vinda, which was a large proglacial river in the past (during the Little Ice Age) and declined due to the Bungebreen glacier's recession. *Photo* J. Dudek, 2008

especially if they are located on steep slopes and in significantly inclined upper parts of valleys. However, some of them, especially at the foot of the slope and in the lower parts of valleys, are being covered by new deposits, first of all moraine or glacifluvial deposits, but also talus, weathering, and fluvial deposits.

Marginal zones began to appear in the front and around the lowest parts of glacier tongues in the beginning of the twentieth century, when the glaciers were the widest (Maps 1a–1b). These zones have been expanding due to frontal and lateral recession of glacier tongues ever since. Marginal zones develop from single ice-cored frontal moraine ridges into extensive complexes of different landforms, including intramarginal sandurs with proglacial river beds, kettle holes, kames, eskers, and low and narrow ridges of fluted moraines. Landforms of one type are transformed into landforms of another type, for example, ice-cored moraine ridges into undulated moraine plains.

A particularly clear change was visible in the valley above the lowest part of the Gåsbreen glacier due to its shortening in the period 2001–2004. The ice-dammed lake (Goësvatnet) was still there in 2000. However, in 2005, the lake did not exist, and its trough was in an initial stage of fluvial transformation (Ziaja and Ostafin 2007). By 2008, the terraced riverbed built of alluvial deposits was formed, with waterfalls on rock outcrops (Figs. 4.3 and 4.5).

Landforms clearly stabilize in places where dead ice was never present or had thawed completely, which is expressed by a large decrease in the rate of landform evolution. This stabilization occurs in some of the smallest marginal zones and along edges of larger ones. However, dead ice is covered by superficial deposits in the majority of the marginal zones' area. The slow thawing of dead ice causes continuous geomorphic transformation and the lowering of marginal zones. The thawing of ice is stimulated by the influx of freshwater and seawater. This leads to the appearance of characteristic ephemeral forms of thaw holes and tongues of mud-flows (Fig. 4.8).

Of course, the cliffs of tidewater glaciers undergo the quickest changes, that is, retreat (Fig. 4.4). The Stormbukta bay as a whole appeared due to the recession of Olsokbreen since 1900 (Wassiliew 1925). The new shore of this large bay is continuously transformed. The sea took 750 m of coastline and 11 ha of land along the northern shore of the Stormbukta bay, and removed the coastline inland a distance of 5–250 m, mainly due to ablation of dead ice (Figs. 4.4 and 4.9) from 1990 to 2008. This process continues, which is evidenced by mud-flows from the lateral ice-cored moraine to the sea (Fig. 4.8). However, at a shorter distance to the glacier front, the land has widened by about 100–200 m (9 ha) due to marine accumulation of the former moraine and glacifluvial material from 1990 to 2008 (Map 1b, Figs. 4.4 and 4.10). The southern Stormbukta coast experiences the same processes. On the eastern coast of the small bay at the front of the Körberbreen glacier (Map 1a) 2.3 ha were added to the land in a similar way.

The geomorphic transformation of areas devoid of glaciers in a warming climate is much slower (than described above) but noticeable. Changes in some parts of the shoreline are the most visible. Geomorphic sea action along the coast increased

Fig. 4.8 Mudflow due to ablation of the dead-ice core of the Olsokbreen lateral moraine, on the northern Stormbukta coast. *Photo* J. Niedźwiecki, 2008

Fig. 4.9 New northern coast of Stormbukta formed after the retreat of the Olsokbreen glacier since the 1980s, due to ablation of dead-ice cores in the lateral moraine and sea action. *Photo* J. Niedźwiecki, 2008

Fig. 4.10 New northern coast of Stormbukta formed after the retreat of the Olsokbreen glacier since the 1990s, due to sea action. *Photo* J. Niedźwiecki, 2008

Fig. 4.11 Results of intensified geomorphic sea action in western Sørkapp Land during a few recent decades: the accumulation coast south of the Palffyodden headland. View to the north. *Photo* J. Dudek, 2008

considerably (especially during storms) due to a significant shortening of the sea-ice season since the 1980s. Sea accumulation prevails in the bay south of the Palffyodden headland: the beach widens at a rate up to 1 m per year (Fig. 4.11). The reverse process, sea abrasion, acts on the Tørrflya plain built of Old-Holocene marine deposits: the beach coastline retreats up to 1 m per year, and the cliff behind the beach (cut from above during storms only), up to 2 m per year (Fig. 4.12).

The reindeer population, regenerated in the 1990s, is a new geomorphic factor affecting large areas of western Sørkapp Land. There are simply too many reindeer, about 170 in 2008, because they are not threatened by any predators. They destroy vegetation and initiate or intensify erosion processes in many places by overgrazing, for example, river erosion in the Hohenloheskardet valley and wind erosion on the Tørrflya (Fig. 4.13) and Bjørnbeinflya plains. Small dunes, which were overgrown and stabilized before (Gębica and Szczęsny 1988), have been made mobile by reindeer. Reindeer also impact microrelief on gentle slopes overgrown by tundra by trampling and thus forming so-called cattle terraces, such as at the foot of Sergeijevfjellet from the sea-side (Fig. 3.4).

Karst processes have intensified due to higher air temperatures and larger quantities of flowing water. Rapid development of karst holes was observed on the Olsokflya lowland. The holes were no deeper than 1 m during the summer of 1986. Some of the holes were 1.5–2.0 m deep in 2000, and at least 10–20 cm deeper in 2008. The enlargement of one of them resulted in the collapse of a cave

Fig. 4.12 Results of intensified geomorphic sea action in western Sørkapp Land during a few recent decades: the abrasion coast of the Tørrflya plain, approximately 9 m. high, with an uncovered whalebone, on the *right*. *Photo* J. Niedźwiecki, 2008

Fig. 4.13 Small dunes that are a result of damage to the vegetation cover due to overgrazing and trampling by reindeer on the Tørrflya coastal plain. *Photo* J. Niedźwiecki, 2008

(karst channel) roof just before Trollosen Spring, which is a pro-glacial river (Fig. 3.3). The development of similar karst holes may also be observed above this karst channel.

Smaller changes in landforms include debris-flows on talus and talus-torrent fans as well as a halt in the development of nivation moraines.

4.4   Changes in Water Drainage Networks

Wiesław Ziaja and Krzysztof Ostafin

Climate warming has led to an increase in the quantity of liquid water in the Sørkapp Land natural environment due to more intensive ablation of snow patches and glaciers, and the deepening of the active layer (with ice) above permafrost.

This water flows to the sea in streams and rivers (up to 10 km long), under the ground in the active layer of the Quaternary deposits, and under the ground through rock fissures and karst channels. Surface and underground streams carry much more water today than in the 1980s. However, the soil and the surface are generally drier in the lowland areas between the streams today due to the deepening of the active layer above the permafrost.

The river and lake network changed the most due to glacier recession.

The ice-dammed Goësvatnet lake disappeared due to the recession of the Gåsbreen glacier. Its water table became lower and its surface decreased due to the thinning of the glacier tongue, which dammed the valley over time. However, the glacier did not become shorter and the lake persisted until 2000. After 2000, recession further shortened the glacier, removing the natural glacial dam, as detected in 2005 (Ziaja and Ostafin 2007). A large river flows from the upper part of the valley through the former lake's trough and is supplied below by water from the Gåsbreen glacier (Figs. 4.3 and 4.5). A few smaller ice-dammed lakes disappeared as well.

However, other lakes have become larger due to glacier recession. This is especially true of a large lake dammed by a frontal moraine in the eastern part of the Bungebreen glacier's marginal zone. A lot of smaller lakes appear, change their shape, and disappear because of dead ice thawing in the larger glaciers' marginal zones.

Pro-glacial rivers change their course and flow both in the marginal zones of glaciers and their forefields, that is, on the coastal lowlands. The greatest changes have occurred in the Bungebreen glacier fore-field. The previously large and rather short Vinda river has virtually disappeared due to the recession of the Bungebreen glacier. Today, only small streams flow periodically in its large and deeply incised bed (Fig. 4.7). The Bungeelva river, which flows out of the aforementioned enlarged lake, became clearly larger, at least in summer.

4.5 Changes in Vegetation

Michał Węgrzyn, Maja Lisowska, Maria Olech and Piotr Osyczka

4.5.1 Plant Communities in 1982–1985

The making of phytosociological maps of Arctic areas is difficult because of incomplete knowledge about tundra plant communities. One of the problems is mosaic vegetation made up of many different communities occupying small areas. Available vegetation maps usually show the distribution of physiognomic types of tundra (CAVM Team 2003) or the distribution of plant communities on a large scale (Elvebakk 2005).

A detailed vegetation map catalogues plant communities in a given area, and shows their extent and geographic location (Maps 3a–3b). The phytosociological map of northwest Sørkapp Land (Dubiel and Olech 1991), which was the basis for the current comparative study, was made on a 1:25,000 scale. Mapping was preceded by making a series of relevés according to the classic Braun–Blanquet method and creating phytosociological tables, which enabled the identification of plant communities. Detailed phytosociological research resulted in the selection of 28 plant communities that were clearly distinguished phytosociologically and were visible in the field. The aim of the research was not to define the precise syntax-onomic position and names of the communities in question. The results included a vegetation map, which would be the basis for further comparative studies on vegetation changes in the area and the characterization of the plant communities identified during the research.

A group of plant communities associated with dry habitats was identified based on relief and water conditions. This group includes vegetation types for which high water content in the substratum is not crucial for existence and which are relatively resistant to the drying properties of wind. This type of vegetation is dominated by bryophytes and lichens, which usually obtain water directly from the atmosphere. On the Kulmstranda and Horsundneset plains, large areas of boulders and pebbles made of siliceous sandstones can be found. Most of them were covered by com-munities of epilithic lichens. The *Orphiniospora moriopsis* community occurred in places exposed to strong winds. A licheno-bryophytic community, represented by *Sphaerophorus globosus*, developed between stones. The floristically poor *Racomitrium lanuginosum* community was also found in the region in places cov-ered with small rocks. This community was found quite often in depressions between old storm ridges. The spatial formation of the dominant moss species *R. lanuginosum* followed the direction of local foehn winds (Fig. 4.14). In places that are protected from the wind, *Umbilicaria arctica* and *Umbilicaria cylindrica* com-munities achieved their optimum. In areas where the wind coming from Hornsund is halted by the slopes of Hohenlohefjellet, a moss and lichen-dominated vegetation complex occurred, composed of *O. moriopsis* and *R. lanuginosum* communities.

Much closer to the shoreline, on marine terraces built mainly of sea pebbles, a mosaic of *Catrariella delisei* and *Gymnomitrion coralloides* communities (Fig. 4.15), which usually exist together, could be found. The *C. delisei* community in this formation was dominant in the southern part of the research area, whereas the *G. coralloides* community was found farther north. Patches of this vegetation complex had a characteristic brown-greyish colour. In the central part of Horsundneset, full of rocky ledges near lakes, another type of vegetation mosaic, composed of *C. delisei, G. coralloides*, and *R. lanuginosum* communities, was identified.

Another mosaic dominated by the *C. delisei* community was one composed of *C. delisei, G. coralloides*, and *O. moriopsis* communities. This type of formation, occupying mainly rocky ledges near the sea in an area of strong winds, was rela-tively rare.

The most spectacular vegetation type in the study area was most likely the *Flavocetraria nivalis–Cladonia rangiferina* community. The principal component of this community included numerous species of fruticose epigeic and epi-bryophytic lichens, along with bryophytes (Fig. 4.16). The *F. nivalis–C. rangi-ferina* community could be found on initial soils in areas protected from the wind by mountain slopes (Fig. 4.17). The best-developed patches of this community were noted at the foot of Hohenlohefjellet, whereas a less-developed form was found in the central part of the coastal plain near the sea and lakes. The *F. nivalis–C. rangiferina* community, the *C. delisei* community, and the *G. coralloides* com-munity form a complex found on rocky ledges between lakes.

Fig. 4.14 *Racomitrium lanuginosum* community: a typical shape formed by foehn winds. *Photo* M. Węgrzyn, 2008

Fig. 4.15 A mosaic composed of the *Cetrariella delisei* community and the *Gymnomitrion coralloides* community on the Hornsundneset coastal plain (with the Sergeijevfjellet and Lidfjellet mountain massifs in the background). *Photo* J. Dudek, 2008

Fig. 4.16 General view of tundra vegetation in the central part of western Sørkapp Land in 1982. In the foreground: the *Carex subspathacea* community; in the background: the *Flavocetraria nivalis–Cladonia rangiferina* community with visible patches of *yellow* fruticose lichens. *Photo* M. Olech

Fig. 4.17 *Flavocetraria nivalis–Cladonia rangiferina* community. *Photo* M. Olech, 1982

All the vegetation types described above occupied the northern part of the research area. Farther south, a larger share of vascular plants was observed. Vascular-plant–dominated communities were the most floristically rich in the study area. Their distribution was associated with areas of moderate soil humidity.

The *Saxifraga aizoides* community, one of the most floristically rich communities in the study area, occupied large parts of the Breinesflya plain. This plain is annually flooded by water from melting snow. Therefore, a large amount of water is available, but only at the beginning of the vegetation season. Stony soil in this area includes a large amount of humus and a lot of calcium carbonate.

As in the previous plant community, the *Bistorta vivipara* community also was floristically rich. It developed in the lower parts of the slopes of Sergeijevfjellet and Lidfjellet, as well as on elevated marine terraces, which are very favourable for plant development. The *B. vivipara* community also formed a species-poor complex of vegetation with the *C. delisei* and the *G. coralloides* communities. This complex was found at numerous locations on marine terraces of Breinesflya and on the slopes of Kvartsittrabben.

The *Saxifraga nivalis* community was often found on moist, western mountain slopes. The substratum in these areas was fine-grained with signs of solifluction. This community was rich in species of both vascular plants and cryptogams.

The *Papaver dahlianum* community developed on scree slopes with western exposure, and on the gravelly sides of seasonal streams (Fig. 4.18). This community was floristically poor, mainly due to unstable ground.

The *Juncus biglumis* community was dominant in the southern part of the research area, that is, on the Breinesflya plain and on the slopes of Lidfjellet and Wiederfjellet. Large patches of this community were found in places with long-lasting snow cover and frequent solifluction episodes. The ground in these areas is moist during the entire vegetation season.

Rather infrequent in the study area, the *Saxifraga oppositifolia* community occupied seashore areas with poor gravelly-sandy soil.

Another group of plant communities was associated with wet areas such as land depressions, stream beds, and lake shores. They were generally found in the central and southern parts of the research area.

A mosaic composed of the *Dupontia pelligera* community and the *Arctophila fulva* community appeared along streams and on moist lake shores. The first community preferred a very wet habitat whereas the second grew in less moist places. This complex included numerous graminoids and resembled meadow formations.

The *Carex subspathacea* community developed in the wettest areas, often flooded during the snowmelt season, and also by permanent and seasonal streams (Fig. 4.16). Under a layer of vascular plants and bryophytes, a 15-cm thick layer of peat was formed on gravelly or gravelly-sandy substratum. The largest areas of the *C. subspathacea* community were found usually far from the sea, in land depressions at the foot of Lidfjellet and Sergeijevfjellet.

A typical peat-producing vegetation type in Sørkapp Land was the *Saxifraga hyperborea–Ranunclus spitsbergensis* community. This very floristically poor community formed in boggy areas on silt or sand with the addition of gravel. The

Fig. 4.18 *Papaver dahlianum* with very expansive *Salix polaris* in the *Papaver dahlianum* community. *Photo* M. Węgrzyn, 2008

layer of peat underneath the plants was 20–80 cm thick. Vascular plants typical for this community occurred infrequently within thick patches of mosses, mainly *Calliergon stramineum*.

Gravelly areas close to streams running across coastal plains were occupied by a very species-poor, moss-dominated *C. stramineum* community. During the summer, areas occupied by these communities are often flooded by the changing water level of the streams. In the tundra landscape, one can easily spot those deep green patches among vegetation in dry habitats, which is grey or in various shades of brown (Fig. 4.16). In larger land depressions, a similar vegetation type can be found, that is, the *Calliergon sarmentosum* community.

Ornithocoprophilic vegetation, associated with seabird colonies, is completely different from the previous types. Soil constantly fertilized by guano rich in nutrients, that is, compounds of phosphorus and nitrogen, is the key habitat-creating factor. In the study area, large nesting colonies of the Little Auk, *Alle alle*, found on the slopes of Hohenlohefjellet and Sergeijevfjellet fertilize the nearby area.

Ornithocoprophilic vegetation, which developed directly in bird colony areas included the *Candelariella arctica* community, composed of epilithic lichens with predominant yellow-coloured *C. arctica*. It developed on the surfaces of rocks and boulders of the Hohenlohefjellet and Sergeijevfjellet slopes.

The *Tetraplodon mnioides* community, on the other hand, developed below bird colonies. It benefited from phosphorus and nitrogen washed out from bird colonies and carried by water into streams. This community, composed entirely of the ornithocoprophilic bryophytes *Tetraplodon mnioides* and *Aplodon wormskjoldii*, was found at the base of the mountains where the soil was soppy.

Apart from all the plant formations mentioned above, one could also find small areas almost entirely devoid of vegetation, where only initial communities of epilithic lichens have developed. These places were located in high mountain areas covered by snow most of the year.

4.5.2 Plant Communities in 2008—Changes Over the Last 25 Years

The analysis of vegetation in Sørkapp Land done in 1982 and 1985 (Dubiel and Olech 1990, 1991) was repeated in 2008. Historical phytosociological data and maps were compared with new results (Maps 3a–3b and 4a–4b).

Phytosociological methodology is very effective for monitoring of the state of vegetation and for observing changes in the composition and distribution of different vegetation types over time. Relevés made in the same area as during the previous study (Dubiel and Olech 1990, 1991) and a repeat of vegetation mapping allowed for an analysis of changes in vegetation in the area.

Vegetation dynamics in the research area and trends, in the context of factors operating during the last 25 years and currently, were analysed.

On the basis of repeated vegetation mapping, significant changes in composition and extent of several plant communities were documented. The principal trend is a decrease in species diversity, leading to a more uniform vegetation. This process is most visible in dry habitats, whereas in wet areas, changes are less noticeable.

One of the most spectacular changes was observed in the complete degradation of the *F. nivalis–C. rangiferina* community at the foot of Hohenlohefjellet and Sergeijevfjellet. This community, previously dominated by fruticose lichens from the genera *Cladonia* and *Cetraria*, has completely transformed. Currently,

F. nivalis, *C. rangiferina*, and other species of *Cladonia* have practically disappeared. The moss *R. lanuginosum* and the lichen *S. globosus* became dominant instead. A larger number of vascular plants, especially *Luzula arcuata*, were also recorded. It seems that the *F. nivalis–C. rangiferina* community transformed into a moss-dominated one, with a predominance of *R. lanuginosum* (Fig. 4.19).

The main component of the *F. nivalis–C. rangiferina* community, fruticose lichens, was almost entirely eliminated by the reindeer. The remaining lichens gradually declined due to the fertilization of the substratum. Currently, these lichen species can only be found on very steep slopes of Hohenlohefjellet and in rock crevices on coastal plains, in places inaccessible to reindeer.

The patches of the *F. nivalis–C. rangiferina* community, which used to form a mosaic together with the *C. delisei* and the *G. coralloides* communities, were also strongly transformed. Currently, only *C. delisei* and *G. coralloides* are present in these areas.

A similar trend of unification is visible in places where a mosaic of the *R. lanuginosum* and the *O. moriopsis* communities used to exist. Presently, boundaries between these two communities have become unclear as a result of the expansion of *R. lanuginosum*. This species used to occupy depressions between storm ridges, where fine-grained rock material would accumulate. Now its extent has moved onto the top parts of the ridges.

Within vegetation complexes with a predominance of the *C. delisei* community, no clear boundaries between the complex components currently exist. *C. delisei* has expanded greatly along the shore, just as *R. lanuginosum* has done inland.

The *C. delisei* community creates a well-defined complex with the *G. coralloides* community. Within this mosaic, the expansion of the *C. delisei* community is also visible.

In the central part of the Horsundneset plain, an earlier map showed a complex of the *C. delisei*, the *G. coralloides*, and *R. lanuginosum* communities, however, currently the *C. delisei* community has expanded. Boundaries between the components have vanished, which reflects the general trend in the area, and the same situation is visible in the complex of the *C. delisei* and the *O. moriopsis* communities. In the latter complex, stony ledges occupied previously by *O. moriopsis* and other epilithic lichens, are comparatively small patches in an area becoming more and more dominated by fruticose lichens.

Changes in the structure of plant communities in wet areas are not as substantial as in communities associated with dry habitats.

The structure of the *Saxifraga aizoides* community, which covers large areas of the Breinesflya plain, has not changed since the 1980s.

In the *B. vivipara* community, which is also floristically rich, an increase in abundance of *Salix polaris* was recorded. It is clear that *S. polaris* is an expansive species in this community.

On the shore, in the *C. delisei* and the *G. coralloides* communities, expansion of *S. polaris* was not observed. Consequently, the expansion of this species may be directed from the shore inland, towards mountain slopes.

(a)

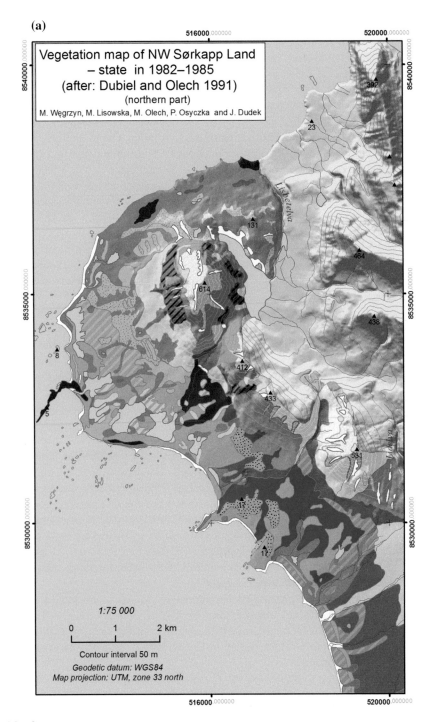

Vegetation map of NW Sørkapp Land
– state in 1982–1985
(after: Dubiel and Olech 1991)
(northern part)
M. Węgrzyn, M. Lisowska, M. Olech, P. Osyczka and J. Dudek

1:75 000

0   1   2 km

Contour interval 50 m
Geodetic datum: WGS84
Map projection: UTM, zone 33 north

Map 3a

(b)

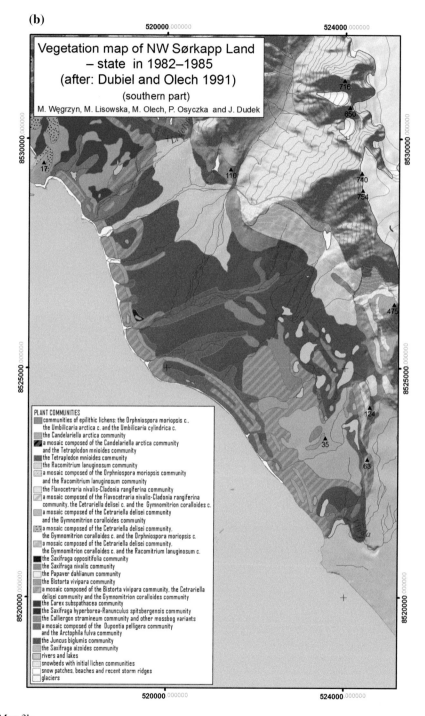

Map 3b

(a)

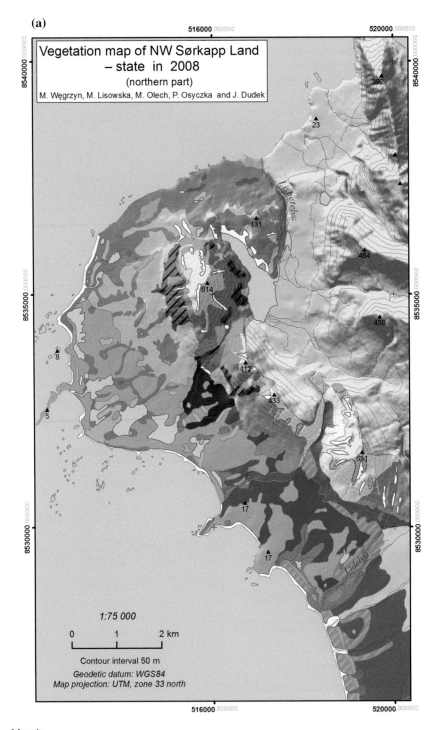

Map 4a

(b)

Map 4b

Fig. 4.19 Degradated form of the *Flavocetraria nivalis–Cladonia rangiferina* community as a result of reindeer grazing, trampling, and fertilizing over the last few decades. *Photo* M. Lisowska, 2008

The dynamics of the *S. nivalis* community is stable, and no changes in the structure of the community were observed. However, within the *P. dahlianum* community, which occupies the western slopes of the mountains, expansion of *S. polaris* is evident (Fig. 4.18). In the 1980s, *S. polaris* used to be found in this community only occasionally. However, now it grows in dense patches.

Occurrence of *Luzula arcuata* was noted in this community for the first time.

Within the *J. biglumis* community, which dominates in the southern parts of the research area, on the Breinesflya plain and at the foot of Lidfijellet and Wiederfjellet, the structure remains unchanged. Both the boundaries of the formation and the species structure seem to be stable.

The *S. oppositifolia* community was very diverse in the 1980s. Presently, major changes can be observed within this pioneer community. The previously dominant *S. oppositifolia* is in decline, whereas other expansive species, which used to be scarce, appear in larger quantities. *S. polaris* is found in large quantities, as is *C. delisei*, which migrates from neighbouring communities. *G. coralloides* can be observed to a smaller extent as well. On the updated vegetation map, the *S. oppositifolia* community includes a mosaic composed of the *C. delisei* and the *G. coralloides* communities.

Fig. 4.20 *Saxifraga hyperborea–Ranunculus spitsbergensis* community. *Photo* M. Węgrzyn, 2008

In permanently wet areas, the extent of the following communities did not change: the *Dupontia pelligera* and the *A. fulva* communities, the *C. subspathacea* community, the *S. hyperborea–Ranunculus spitsbergensis* community (Fig. 4.20), the *C. stramineum* community, and other moss bog variants. Species composition also does not fundamentally deviate from the one noted in the 1980s. Only the general condition of plants and the state of preservation have visibly changed due to reindeer grazing.

Within ornithocoprophilic vegetation, species composition remained the same and changes in their extent were not observed. This is associated with the specific character of these communities, one being composed of epilithic lichens, with leading species *C. arctica*, and the other typically bryophytic, created mainly by *Tetraplodon mnioides*. The size of the colonies of *A. alle* seems to be unchanged, therefore ornithocoprophilic vegetation constantly receives enough nutrients to prevail.

Communities of epilithic lichens, with the *U. arctica* and the *U. cylindrica* communities, are rather stable and have not transformed much over the last 25 years. Patches of these communities are found in the northern part of the study area, where extensive rock surfaces, boulders, and storm ridges made of sea pebbles occur frequently.

4.5.3   Causes of Changes

Vegetation maps are a valuable source of information for estimating changes in vegetation in a given area. They are also very useful for assessing the speed and direction of changes, which is particularly important when vegetation is dynamic, influenced by numerous factors such as climate changes and animal and human impact.

These factors and the changes caused by them have been thoroughly analysed throughout the Arctic. The research results provide insight into the speed of the changes and their direction, and allow us to make predictions concerning future trends.

Svalbard ecosystems are described in the literature as adapted to extreme fluctuations in climate conditions on different temporal scales and slowly reacting to climate changes (Jónsdóttir 2005). Other research confirms this, revealing that the impact of climate change on vegetation cannot be detected in Svalbard. Vegetation mapping in Adolfbukta (Billefjorden, middle Spitsbergen) repeated in 2008 after 70 years did not show any changes in vegetation (Prach et al. 2010). The authors suggest that the reason for such a state of affairs, apart from the Arctic ecosystems' slow reaction to climate changes, could be the pressure of herbivores, masking the possible development of vegetation in that area.

In the case of Sørkapp Land, research results show a different situation. Major changes in vegetation have taken place since the 1980s, especially in dry or slightly moist areas. On the contrary, minute changes were recorded in very wet areas. The results of experimental research in Central Spitsbergen (Speed et al. 2010) were in line with the observations from Sørkapp Land. An herbivory simulation (goose pressure) showed that different vegetation types have different levels of resistance to herbivore disturbance. Wetlands and mires proved to be the most resistant.

The main cause of vegetation changes in Sørkapp Land is the rapidly growing population of reindeer in the area. In addition to environmental changes favouring reindeer expansion, the legal protection of these animals in the area was also of great importance thanks to the creation of South Spitsbergen National Park in 1973. A rapid expansion of the reindeer population was observed, from 1 or 2 animals in the 1980s to about 100 after the year 2000 (Ziaja 2002), and later to about 170 animals in 2008. The growing number of reindeer has had a strong effect on the structure and distribution of plant communities. The impact of herbivores has included grazing on lichens and over-ground parts of vascular plants, as well as the trampling of bogs (Fig. 4.21) and the fertilisation of the ground.

Svalbard reindeer, unlike other reindeer subspecies, rarely migrate (Tyler and Øritsland 1989). It can thus be assumed, that in the future the population of reindeer in western Sørkapp Land will not decrease, provided there is enough forage.

In comparison, reintroduction of reindeer in the area of Ny Ålesund, Northwest Spitsbergen, caused a major transformation of the tundra over a short period of time, and the effects are similar to those observed in Sørkapp Land. Twelve reindeer were brought in 1978 to Brøggerhalvøya, near Ny Ålesund. Until 1986–1987, the

Fig. 4.21 Wet moss tundra in the outlet of the Hohenloheskardet valley at a time without any reindeer in the *upper photograph*, and the same tundra with a visible effect of reindeer grazing and trampling since the 1990s in the *bottom photo*. *Top photo* W. Ziaja. *Bottom photo*: M. Węgrzyn

number of animals increased to 110 (Øritsland 1987). In 1991 there were already 300 reindeer (Wegner et al. 1992, Elvebakk 1997). Only one year after the introduction of reindeer, the destruction of the *F. nivalis* population in Brøggerhalvøya was observed, and one year after that, *Cladonia mitis* [syn. *C. arbuscula* ssp. *mitis*] disappeared as well. After some time, it was observed that a part of the *C. delisei* population, previously not grazed on by reindeer, was also destroyed (Elvebakk 1997). Negative changes in the population of epigeic fruticose lichens can also be observed in other regions of the Arctic, for instance, in Alaska (Joly et al. 2009), where a decrease in cover and biomass of ground lichens was recorded in recent years. This tendency is associated with several factors such as reindeer and caribou activity as well as competition with vascular plants spreading as a result of climate change. An experiment simulating the influence of global warming on tundra resulted in a decrease in lichen cover (Hudson and Henry 2010).

References

Brázdil R (1988) Variation of air temperature and atmospheric precipitation in the region of Svalbard. In: Results of investing of the Geogr. Research expedition Spitsbergen 1985. Folia Facultatis Scientiarum Naturalium Universitatis Purkynianae Brunensis (Brno), Geographia 24:285–323

CAVM Team (2003) Circumpolar arctic vegetation map. Scale 1:7,500,000. Conservation of Arctic Flora and Fauna (CAFF) Map No. 1. U.S. Fish and Wildlife Service, Anchorage, Alaska

Dubiel E, Olech M (1990) Plant Communities of NW Sörkapp Land (Spitsbergen). Zeszyty Naukowe Uniwersytetu Jagiellońskiego, Prace Botaniczne 21:35–74

Dubiel E, Olech M (1991) Phytosociological map of NW Sörkapp Land (Spitsbergen). Zeszyty Naukowe Uniwersytetu Jagiellońskiego, Prace Botaniczne 22:47–54

Elvebakk A (1997) Tundra diversity and ecological characteristics of Svalbard. In: Wielgolaski FE (ed) Polar and alpine tundra. Ecosystems of the world, vol 3. Elsevier, Amsterdam, pp 347–359

Elvebakk A (2005) A vegetation map of Svalbard on the scale 1:3,5 mill. Phytocoenologia 35 (4):951–967

Gębica P, Szczęsny R (1988) Symptoms of aeolian accumulation in western Sørkapp Land, Spitsbergen. Pol Polar Res 9(4):447–460

Hudson JMG, Henry GHR (2010) High arctic plant community resist 15 years of experimental warming. J Ecol 98:1035–1041

Joly K, Jandt RR, Klein DR (2009) Decrease of lichens in arctic ecosystems: the role of wildfire, caribou, reindeer, competition and climate in north-western Alaska. Polar Res 28:433–442

Jónsdóttir IS (2005) Terrestrial ecosystems on svalbard: heterogeneity, complexity and fragility from an Arctic Island perspective. Biol Environ Proc R Irish Acad 105(3):155–165

Łupikasza E (2013) 11. Atmospheric precipitation. In: Marsz AA, Styszyńska A (eds) Climate and climate change at Hornsund, Svalbard. Gdynia Maritime University, Gdynia, p 199–211

Marsz AA (2013) 9.1. Annual air temperature. 9.2. Monthly air temperatures. 9.3. The annual patterns of diurnal temperature. 9.4. Thermal seasons. In: Marsz AA, Styszyńska A (eds) Climate and climate change at Hornsund, Svalbard. Gdynia Maritime University, Gdynia, p 145–159

Marsz AA, Styszyńska A (eds) (2013) Climate and climate change at Hornsund, Svalbard. Gdynia Maritime Uniwersity, Gdynia, 402 p

Øritsland NA (1987) The reindeer in the Ny   lesund area. Norsk Polar Institutt   rbok 1986–1987:38–40

Prach K, Košnar J, Klimešová J, Hais M (2010) High Arctic vegetation after 70 years: a repeated analysis from Svalbard. Polar Biol 33(5):635–639

Pillewizer W (1939) Die Kartographischen und Gletscherkundlichen Ergebnisse der Deutschen Spitsbergenexpedition 1938. Petermanns Geographische Mitteilungen, vol 238, 46 p +2 maps

Speed JDM, Cooper EJ, Jónsdóttir IS, van der Wal R, Woodin SJ (2010) Plant community properties predict vegetation resilience to herbivore disturbance in the Arctic. J Ecol 98:1002–1013

Styszyńska A (2013) Results of observations. In: Marsz AA, Styszyńska A (eds) Climate and climate change at Hornsund, Svalbard. Gdynia Maritime University, Gdynia, p 321–365

Sund M, Eiken T, Hagen JO, Kääb A (2009) Svalbard surge dynamics derived from geometric changes. Ann Glaciol 50(52):50–60

Tyler NJC, Øritsland NA (1989) Why don't Svalbard reindeer migrate? Ecography 12(4):369–376

Wassiliew (Vasiliev) AS (1925) Spitsberg III. Feuille du Sud. (map). In: Missions scientifiques pour la mesure d'un arc de méridien au Spitsberg. Miss. Russe, Publication de l'Académie des Scientes de Russie

Wegner C, Hansen M, Jacobsen LB (1992) Vegetasjonsovervåkning på Svalbard 1991. Effekter av reinbeite ved Kongsfjorden, Svalbard. Norsk Polarinstitutt Meddelelser 121, 54 p

Ziaja W (1988) Fizycznogeograficzne zróżnicowanie górskiej części północno-zachodniego Sørkapplandu (Spitsbergen). Doctoral thesis (manuscript), Jagiellonian University, Institute of Geography, 137 p

Ziaja W (1999) Rozwój geosystemu Sørkapplandu, Svalbard. Wydawnictwo Uniwersytetu Jagiellońskiego, Kraków, 105 p

Ziaja W (2002) Changes in the landscape structure of Sørkapp Land. In: Ziaja W, Skiba S (eds) Sørkapp Land landscape structure and functioning. Wydawnictwo Uniwersytetu Jagiellońskiego, Kraków, p 18–50

Ziaja W (2004) Spitsbergen Landscape under 20th century climate change: Sørkapp Land. Ambio 33(6):295–299

Ziaja W, Ostafin K (2007) Współczesna przemiana krajobrazu Lodowca Gås i okolicy. In: Przybylak R, Kejna M, Araźny A, Głowacki P (eds) Abiotyczne środowisko Spitsbergenu w latach 2005–2006 w warunkach globalnego ocieplenia. Uniwersytet Mikołaja Kopernika, Toruń, p 235–245

Chapter 5
Conclusions and Prognosis
for Environmental Change

**Wiesław Ziaja, Michał Węgrzyn, Maja Lisowska, Maria Olech
and Piotr Osyczka**

Abstract The first and direct result of climate warming has been glacial recession, which stimulated an entire process of landscape (and seascape) changes along the eastern boundary between western Sørkapp Land (devoid of glaciers) and the glaciated peninsula interior. A completely new landscape has appeared there. Also fore-fields of glaciers have been indirectly influenced by the glaciers' retreat. Some sequences of non-glacial and non-postglacial coastline have been affected by an increase in the geomorphic activity of the sea due to a shorter sea-ice season. During the next few decades, the described trend of environmental-landscape transformation will continue unless the climate cools down. In the case of a progressive warming, the extensive tongues of big glaciers will first retreat and then disappear. The main result of that would be an expansion of non-glacial landscape, vegetation and animal life to the east, into the currently glaciated peninsula's interior. On the basis of repeated vegetation mapping, significant changes in composition and extent of several plant communities were documented. Decrease in species diversity, leading to a more uniform vegetation, has been observed mainly in dry habitats. In some cases boundaries between plant communities that were clear in the 1980s, have now vanished. Fruticose epigeic lichens, like *Flavocetraria nivalis*, *Cladonia rangiferina*, and other species of *Cladonia* have disappeared from the most part of the study area and their extent is now limited to steep rocky slopes. In some communities increase in abundance of *Salix polaris* was recorded. The main cause of vegetation changes in Sørkapp Land is the rapidly growing reindeer population in the area.

Keyword Deglaciation of landscape · Reindeer impact on vegetation

W. Ziaja (✉)
Institute of Geography and Spatial Management, Jagiellonian University, Cracow, Poland
e-mail: wieslaw.ziaja@uj.edu.pl

M. Węgrzyn · M. Olech · P. Osyczka
Department of Polar Research and Documentation, Institute of Botany,
Jagiellonian University, Cracow, Poland

M. Lisowska
Centre for Polar Studies, University of Silesia, Sosnowiec, Poland

© The Author(s) 2016
W. Ziaja (ed.), *Transformation of the Natural Environment in Western
Sørkapp Land (Spitsbergen) since the 1980s*, SpringerBriefs in Geography,
DOI 10.1007/978-3-319-26574-2_5

5.1 Landscape Development

Wiesław Ziaja

A clear climate change, reflected in a significant increase in air temperature and atmospheric precipitation, is the reason for an increase rate of landscape development. Therefore, each environmental-landscape component, apart from solid bedrock, has changed.

The first and direct result of climate warming has been glacial recession, which stimulated an entire process of landscape (and seascape) changes along the eastern boundary between western Sørkapp Land (devoid of glaciers during the Holocene) and the glaciated peninsula interior. These profound landscape changes, described above, have mainly been the result of the after-effects of the glaciers' retreat. A completely new landscape has appeared.

The fore-fields of glaciers, i.e. areas outside the glaciers' maximum extent, have been indirectly influenced by the glaciers' retreat. Landforms, Quaternary deposits and water networks have been significantly affected, especially on the coastal plains in the north and south of the study area: Gåshamnøyra and Tørrflya.

Some sequences of non-glacial and non-postglacial coastline have been affected by an increase in the geomorphic activity of the sea due to a shorter sea-ice season, which can be observed via a clear retreat (under abrasion) or advance (under accumulation) of the coastline. The formation of new coastlines at the front of the tidewater glaciers (Körberbreen and Olsokbreen) is a more complicated process.

Current environmental and landscape changes in large parts of the coastal lowlands and mountains—being ice-free during the Holocene—are much more successive and not as explicit as changes previously mentioned. However, a longer summer season and less snow patch persistence, an increased quantity of water in the ground surface and the active layer of permafrost affected geomorphic and hydrologic processes everywhere. After 25 years, this effect is visible in a slower rise of nivation moraines, more frequent debris-flows and more intensive fluvial processes.

During the next few decades, the described trend of environmental-landscape transformation will continue unless the climate cools down. In the case of a progressive warming, the extensive tongues of the Gåsbreen, Bungebreen and Vitkovskibreen glaciers will first retreat and then disappear. The principal reason for this is that the main part of each glacier is located at an altitude of 300 m above sea level or less. The same is true of low-lying smaller glaciers or their parts. The main result of that would be an expansion of non-glacial landscape, vegetation and animal life to the east, into the currently glaciated peninsula's interior.

5.2 Vegetation Changes

Michał Węgrzyn, Maja Lisowska, Maria Olech and Piotr Osyczka

On the basis of tundra vegetation mapping repeated after 25 years, major changes in the structure of several plant communities and their extent were recorded. First of all, a decrease in species diversity is visible, leading to more homogenous vegetation. This process is more noticeable in dry areas, whereas in wet areas, the changes are smaller. One of the most spectacular changes was a complete transformation of the *Flavocetraria nivalis—Cladonia rangiferina* community, including a strong decline of the fruticose ground lichen populations of this community.

It seems necessary to try to predict the direction of future vegetation changes in Sørkapp Land. It is probable that the reindeer population in western Sørkapp Land will not decrease over the next few years. Also, climate change scenarios indicate a high probability of further warming in the area.

The results of research from other parts of the Arctic can provide necessary information for estimating future changes. It is assumed that in the areas grazed by reindeer, graminoids usually increase (Elvebakk 1997). This is also true in lichen-dominated plant communities (Klein 1968; Post and Klein 1999) and they are also predicted to increase under global warming scenarios (Walker et al. 2006). On the other hand, macrolichens, although important for the functioning and biodiversity of cold northern ecosystems, are predicted to be negatively affected by climate change (Cornelissen et al. 2001). Therefore, the increase of graminoids in the coming years is likely to take place in the research area.

Apart from their main aim, the botanical research in Sørkapp Land from the 1980s (Dubiel and Olech 1990, 1991) captured a state of tundra vegetation in a specific state—developing without any presence of large herbivores for a considerable long period of time (several decades). Data from that time (Dubiel and Olech 1990, 1991) shows distinct characteristics of reindeer-free tundra. However, the vegetation state in the 1980s was most probably not a natural one.

Before that time, in the 19th and the beginning of the 20th century, there was a large reindeer population in Sørkapp Land and the whole Svalbard. Subsequent rapid increase of hunting pressure led to severe decline of the number of reindeer and thus to the development of herbivory-sensitive vegetation communities. Unfortunately, the first botanical records from Svalbard are purely floristic (Nathorst 1883; Resvoll-Holmsen 1927) and no information about plant communities was available from that time. It was not until 1940 when phytosociological research started in this area (Hadač 1946). Since the disappearance of reindeer from Sørkapp Land was associated with human impact, it can be claimed that the tundra vegetation without the impact of reindeer grazing and trampling was not natural for this area.

The degree of preservation of fruticose lichens may be treated as a good indicator of changes taking place in the presence of a large population of reindeer. It can be also, indirectly, used to determine the level of effects of climate change in the Arctic (Cornelissen et al. 2001; Walker et al. 2006; Joly et al. 2009).

References

Cornelissen JHC, Callaghan TV, Alatalo JM, Michelsen A, Graglia E, Hartley AE, Hik DS, Hobbie SE, Press MC, Robinson CH, Henry GHR, Shaver GR, Phoenix GK, Gwynn Jones D, Jonasson S, Chapin FS III, Molau U, Neill C, Lee JA, Melillo JM, Sveinjornsson B, Aerts R (2001) Global change and arctic ecosystems: is the lichen decline a function of increases in vascular plant biomass? J Ecol 89:984–994

Dubiel E, Olech M (1990) Plant communities of NW Sörkapp land (Spitsbergen). Zeszyty Naukowe Uniwersytetu Jagiellońskiego, Prace Botaniczne 21:35–74

Dubiel E, Olech M (1991) Phytosociological map of NW Sörkapp land (Spitsbergen). Zeszyty Naukowe Uniwersytetu Jagiellońskiego, Prace Botaniczne 22:47–54

Elvebakk A (1997) Tundra diversity and ecological characteristics of Svalbard. In: F. E. Wielgolaski (ed) Polar and alpine tundra. Ecosystems of the World, 3. Elsevier, Amsterdam, p 347–359

Hadač E (1946) The plant communities of Sassen Quarter. Vestspitsbergen. Studia botanica Čechoslov. (Praha) 7:127–164

Joly K, Jandt RR, Klein DR (2009) Decrease of lichens in Arctic ecosystems: the role of wildfire, caribou, reindeer, competition and climate in north-western Alaska. Polar Reseach 28:433–442

Klein DR (1968) The introduction, increase, and crash of reindeer on St. Matthew Island. Journal of Wildlife Management 32:350–367

Nathorst AG (1883) Nya bidrag till kännedomen om Spitsbergens kärlväxter, och dess växtgeografiska förhållanden. Kungl. Sv. Vetensk.-Akad. Handl. 20:1–88

Post E, Klein DR (1999) Caribou calf production and seasonal range quality during a population decline. Journal of Wildlife Management 63:335–345

Resvoll-Holmsen H (1927) Svalbards flora. J.V. Cappelena Forlag, Oslo, 56 p

Walker MD, Wahren CH, Hollister RD, Henry GHR, Ahlquist LE, Alatalo JM, Bret-Harte MS, Calef MP, Callaghan TV, Carroll AB, Epstein HE, Jonsdottir IS, Klein JA, Magnusson B, Molau U, Oberbauer SF, Rewa SP, Robinson CH, Shaver GR, Suding KN, Thompson CC, Tolvanen A, Totland Ø, Turner PL, Tweedie CE, Webber PJ, Wookey PA (2006) Plant community responses to experimental warming across the tundra biome. Proceedings of the National Academy of Science 103:1342–1346